# NanoVNAs Explained
## for the radio amateur

by Mike Richards, G4WNC

Radio Society of Great Britain

Published by the Radio Society of Great Britain, 3 Abbey Court, Fraser Road, Priory Business Park, Bedford MK44 3WH. Tel: 01234 832700. Web: www.rsgb.org

First published 2022

© Radio Society of Great Britain, 2022. All rights reserved. No part of this publication may be reproduced, stored in a retrieval system, or transmitted, in any form or by any means, electronic, mechanical, photocopying, recording or otherwise, without the prior written permission of the Radio Society of Great Britain.

ISBN: 9781 9139 9519 5

Cover design: Kevin Williams, M6CYB

Editing, design and layout: Steve Telenius-Lowe, PJ4DX

Production: Mark Allgar, M1MPA

Printed in Great Britain by CPI Anthony Rowe of Chippenham, Wiltshire

Publisher's Note:
The opinions expressed in this book are those of the author(s) and are not necessarily those of the Radio Society of Great Britain. Whilst the information presented is believed to be correct, the author(s), the publishers and their agents cannot accept responsibility for consequences arising from any inaccuracies or omissions.

# Contents

### Part 1

| | |
|---|---|
| Preface | 5 |
| NanoVNA basics | 7 |
| NanoVNA quirks | 9 |
| NanoVNA versions | 13 |
| NanoVNA firmware | 17 |
| The Smith chart | 20 |
| S parameters simplified | 24 |
| Calibration basics | 28 |
| Computer control | 40 |
| NanoVNA and SimSmith | 54 |

### Part 2

| | |
|---|---|
| Introduction | 57 |
| Antennas | 58 |
| ATU settings | 63 |
| Feeder loss | 65 |
| Resonant stubs | 67 |
| Time Domain Reflectometry (TDR) | 71 |
| RF switches and relays | 74 |
| Passive filters | 80 |
| Active filters and amplifiers | 82 |
| Attenuators | 85 |
| Directional couplers | 87 |
| RF tap | 90 |
| Common mode chokes | 92 |
| Baluns | 96 |
| Ununs | 100 |
| Splitter / combiner | 102 |
| Crystals | 104 |
| Cable checker | 106 |
| NanoVNA-H4 menu map | 108 |
| NanoVNA V2 menu map | 109 |

### Index

| | |
|---|---|
| Alphabetical | 110 |
| Subject | 111 |
| Web resources | 112 |

# Preface

Vector Network Analysers have traditionally been out of reach for most radio amateurs, primarily due to cost, but also due to their bulk and designation as a laboratory instrument. This lack of familiarity leaves a knowledge gap where many don't appreciate the value of a VNA when working with RF devices. Other than finding the occasional surplus unit from one of the big manufacturers, the only viable VNA opportunity for most has been the, still excellent, DG8SAQ Vector Network Analyser by SDR-Kits. In recent times there have also been a few projects for home-constructed VNAs and these are often based around the use of an SDR transceiver module.

However, the launch of the NanoVNA brought VNA technology below the £100 mark and has, understandably, generated significant interest. Whilst the availability of a VNA at this price is very attractive, new users face a steep learning curve before they can trust their results. A scan through the NanoVNA user forums will reveal plenty of frustrated users battling to understand the technology. This book has been written with those new users in mind. I've focused on practical guidance and deliberately avoided using complex maths to explain VNA principles. My prime aim in this book is to give the reader sufficient knowledge and guidance to achieve worthwhile results when using any of the popular NanoVNA series.

The book is divided into two parts: Part 1 provides background reading to help you better understand the operation of a VNA and introduces you to many of the essential elements such as Smith charts, S parameters, Calibration, etc. A VNA costing less that £100 will always have compromises, so I've included a section that describes the shortcomings and provides advice on how to mitigate them. I've also provided an analysis of the different NanoVNA hardware models, along with guidance on updating the firmware. Computer control is available for the Nano VNA and I have included a section on how to make the most of the available software.

Part 2 is packed with practical examples of a wide range of VNA-based measurements. This section is specifically intended for new users and those who make occasional use of their VNA. By using the detailed illustrations. combined with step-by-step guides to each measurement type, you will increase your chances of achieving accurate results.

# Part 1: NanoVNA Basics

## NANOVNA BASICS

VNAs (Vector Network Analysers) were first developed in the 1950s and have become standard equipment for any workshop involved in RF design and testing. The VNA is a versatile instrument that can provide detailed performance testing of components, modules and even complete systems Let's take a closer look at the features of the VNA that make it so versatile.

Most of the instruments found in the workshop use scalar measurements, i.e. each measurement produces a single value such as voltage, current, watts, etc. However, these scalar measurements don't tell the full picture when dealing with high frequency signals. This is because every component or connection has a frequency-dependent or reactive element. Even a humble resistor will have a small amount of stray capacitance and inductance. Whilst this may be tiny, it will become increasingly significant as the frequency increases. These stray reactive components can disrupt the expected performance of a device by introducing phase shifts and propagation delays.

To get a more complete view of the high frequency performance of a device or system we need to measure the gain / loss and the complex impedances. To do this accurately, we need to know the amplitude and phase of the input and output signals. However, we also need to check for any reflected signal due to mismatch at the filter input and output. This requires four measurements, as shown in **Fig.1-1**. These four measurements are called *scattering parameters* or *S parameters* and I've covered these in more detail in a separate section.

S parameters provide a very elegant way to describe the RF performance of any device. There

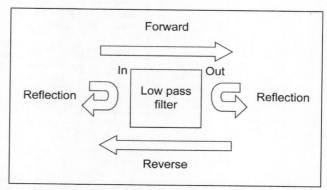

**Fig. 1-1: The four measurements required to classify a network.**

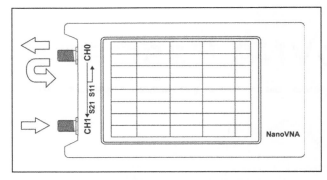

Fig. 1-2: NanoVNA available measurements.

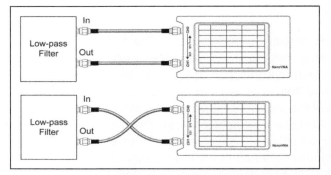

Fig. 1-3: NanoVNA connections for full two-port measurement.

are also standard mathematical formulas available to convert S parameter data into familiar values such as VSWR, gain / loss, return loss, impedance, etc.

Having looked at the basics of a VNA, let's now focus on the *NanoVNA*. Whilst the NanoVNA has two ports, it's not a complete two-port VNA because the second port is receive only, and used solely for through measurements, **Fig.1-2**. However, that is not a significant problem, because we don't often need full two-port measurements. Those cases that do require two-port analysis can be accommodated by reversing the connections to the DUT (Device Under Test), **Fig.1-3**.

In **Fig.1-4** I've shown how the RF signals for CH0 are captured. As you can see, a resistive reflectance bridge is used to extract the reflected values. This is based on a standard Wheatstone bridge and I've shown a more conventional circuit layout in **Fig.1.4**. The resulting RF signals are passed to mixers for down conversion to a 15kHz IF (Intermediate Frequency). The use of such a low IF means

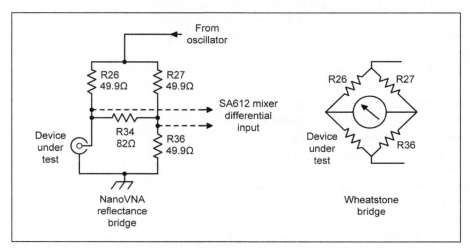

Fig. 1-4: NanoVNA reflectance bridge circuit (left) and redrawn in traditional layout (right).

the signals can be digitised using a low-cost audio soundcard chip. Once digitised, these signals are processed using SDR techniques to extract the amplitude and phase values. The final step is to use a microcontroller to convert the raw signals into familiar values for display. The microcontroller also manages the touchscreen, controls and the USB interface.

Version 2 of the NanoVNA uses the same basic techniques but with a refined design. A few key benefits from Version 2 are:

- Wider frequency – up to 4GHz for some models;
- Improved dynamic range – up to 90dB;
- Uses fundamental stimulus frequencies – eliminates the use of harmonics;
- Improved frequency stability.

## NANOVNA QUIRKS

Given that commercial VNAs cost tens of thousands of pounds, you must expect a few compromises in a device that costs under £100. However, once you understand the shortfalls and their impact on the results, the NanoVNA remains a useful measuring instrument.

The two main limitations of the basic NanoVNA are the restricted measurement points and dynamic range. There are also issues with the use of a square wave test signal and harmonics for the higher frequency range. Here's a description of the issues along with a few ideas of to mitigate their impact.

### Step Limitation

When using the NanoVNA to sweep across a range of frequencies, the stimulus signal changes in discrete steps. In the basic NanoVNA, there are just 101 steps, though later models and updated firmware can extend this to as many as 401 steps. These step limitations apply to calibration as well as to measurements.

There are several ways to manage this limitation. When using a small number of steps, you risk missing any sharp resonances that may be present in the DUT (Device Under Test). A typical example could be a short vertical antenna or small HF loop, where the resonance may be only a few kilohertz wide. If you know that a sharp resonance is likely, you can restrict the NanoVNA's sweep width so that the 101 steps are distributed over a narrower spectrum. An alternative solution, if you have access to a PC, is to use the NanoVNA Saver software. Although originally designed to save S parameter files from the NanoVNA, the current software includes the facility to divide the sweep range into many narrower sweeps called segments. Each segment uses all the available steps, i.e. 101 in a basic NanoVNA. For example, by using ten segments you could expand the steps to 1010. The NanoVNA Saver software manages the segmented sweeps and seamlessly sums the results to provide a combined display. However, the sweep time will increase proportionately with the number of segments. You can read full details in the Computer Control section of this book.

**Dynamic Range**
In the case of the NanoVNA, the dynamic range is the level difference between the *stimulating* or test signal and the noise floor of the measurement system. Professional VNAs would normally have a dynamic range in the order of 120dB or greater, but a basic NanoVNA-H will typically achieve 70dB dynamic range in the 50kHz to 300MHz band, dropping as low as 40dB for the extended ranges.

This still makes the instrument useful, but not for tasks that demand a high dynamic range, such as aligning a repeater diplexer. The main reason for the drop in dynamic range at the higher frequencies is the use of harmonics of the stimulus signal. The output of the Si5351 synthesiser chip is a square wave signal that is rich in odd-order harmonics. The extended frequency coverage of the NanoVNA V1 is achieved by using the 3rd and 5th harmonics of the test signal. Whilst the 3rd and 5th harmonics are relatively strong, the output level available is down by approximately 9.5dB for the 3rd harmonic and 14dB for the 5th harmonic. It's this drop in stimulus level plus noise in the receive circuitry that limits the dynamic range.

In **Fig.1-5** I've shown a spectrum plot of a NanoVNA-H4 set to CW (single frequency) output at 400MHz. The centre spike is the desired signal showing a level of –20dBm. However, to the left is the square wave fundamental at –6dBm. The peak to the right is the 5th harmonic of 133MHz which is at –23dBm. This 5th harmonic is used when measuring above 600MHz.

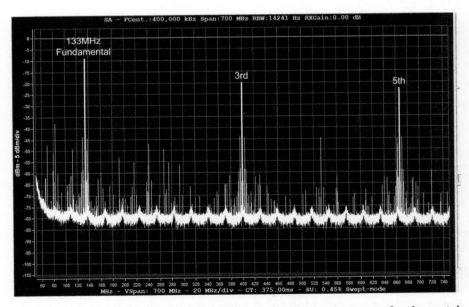

Fig. 1-5: Spectrum analysis of NanoVNA output showing the strong fundamental at 133MHz.

NanoVNA V2 provides a significant improvement in dynamic range thanks to the inclusion of the ADF4350 microwave synthesisers. Although these still produce a square wave output, all measurements are done using just the fundamental. As it's a square wave, there are still plenty of harmonics available but they are not used for measurement.

### Saving to SD Card

Later versions of the NanoVNA-H firmware include a top-level menu item that allows saving the S parameter data to an SD card. However, at the time of writing, most devices don't have an SD card slot in the case! If you look inside the case of your NanoVNA, you will either find a bare PCB space for fitting an SD card socket or a fitted socket, **Fig.1-6**. It is a relatively easy task to make a cutaway in the plastic case to access the SD card slot. The NanoVNA saves the S parameter data in standard Touchstone format ready for transfer to other plotting and analysis tools such as the excellent SimSmith software.

Fig. 1-6: NanoVNA microSD card slot.

### Output Level Adjustment

The original NanoVNA and the newer V2 provide minimal output power control. The NanoVNA-H series gives the user control of the Si5351 output drive level and can be set at 2, 4, 6, 8mA or Auto. The adjustment range is about 10dB, but I suggest you use the Auto option. This attempts to maintain a constant output level by automatically selecting the best mA drive level. Version 2 offers no control of the Si5351 synthesiser output level but does include control of the ADF4350 synthesisers that are used for the higher frequency bands. The adjustment is accessed via the CFG-SWEEP menu item from the STIMULUS menu. The adjustment range is approximately –10dBm to –20dBm in four steps with setting 3 giving the highest output.

### Amplifier Measurements

When using the NanoVNA to measure amplifiers, there are two quirks you need to take into account. The first is the non-adjustable output signal on the NanoVNA V1 that's fixed at about –6dBm for the fundamental and approximately –20dBm and –23dBm for the 3rd and 5th harmonics used above 300MHz. The lack of adjustment could overdrive the amplifier that you're testing. The simple solution is to add an attenuator between the VNA output and the amplifier input. Generally you should use an attenuator value that matches the amplifier's expected gain, i.e. for a 30dB amplifier

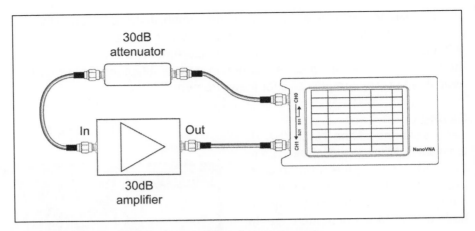

Fig. 1-7: Matching amplifier gain with an input attenuator.

you need to add a 30dB attenuator before the amplifier, **Fig.1-7**. This also helps prevent the amplifier output from overloading the NanoVNA input. The only problem with this solution is that it isolates the amplifier input from the NanoVNA, so you won't be able to measure the input matching.

A second problem occurs when making amplifier measurements above 300MHz. The higher test frequencies are obtained by using the 3rd or 5th harmonic of the square wave from the Si5351. Whilst this works surprisingly well, the lower frequency fundamental will still be present in the test signal and around 14dB higher than the desired signal, **Fig.1-8**. If you're measuring a wideband amplifier, you may need to employ more input attenuation to prevent the fundamental from overdriving the RF amplifier or the NanoVNA input.

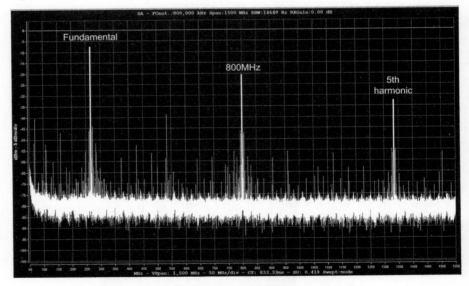

Fig. 1-8: NanoVNA output spectrum for an 800MHz test signal.

## V2 Crystal Measurements

The VNAs can be very helpful in determining the parallel and series resonant characteristics of crystals. However, there is a problem when using a NanoVNA V2 for this purpose. The test signal configuration results in the signal being pulsed at a rate of several kilohertz. Whilst not a problem for other measurements, crystals are electromechanical devices and need settling time after the application of a test signal. The switching frequency of the V2 test signal doesn't leave sufficient settle time, so crystal measurements are likely to be inaccurate. NanoVNA V1 has a continuous test signal so doesn't suffer this problem and is thus a better choice for testing crystals. If you have a NanoVNA V2, there is a special firmware build available specifically for testing crystals. This is available via groups.io in the NanoVNA V2 forum.

## Summary

Despite the shortcomings mentioned here, the NanoVNA remains a popular and useful measurement system. Once you understand its limitations, there are plenty of work-arounds available for common measurements.

# NANOVNA VERSIONS

The original NanoVNA design as shown in **Fig.1-9** was produced several years ago by edy555. Since then there have been numerous copies, clones, a version 2 and a version 3 in the design stage. Whilst the ongoing development keeps the project moving forward, new users can find it hard to

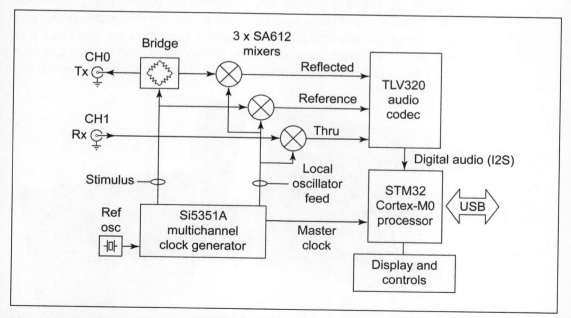

**Fig. 1-9: Original NanoVNA block diagram.**

decide which model to purchase. This is further compounded by model specific firmware releases that add new features. In this section I'll chart the development of the NanoVNA and help you narrow down the choice of model and firmware.

### Original Design

The original NanoVNA design was produced by Edy555 and made available as a kit with all the relevant files stored on Github (**http://github.com/ttrftech/NanoVNA**). I've shown a block diagram of the original design in **Fig.1-9**. Here you can see that a Si5351A programmable clock generator uses a VCTCXO (Voltage Controlled Temperature Compensated Crystal Oscillator) reference to provide the stimulating signal for the measurements as well as the local oscillator feeds for the SA612 active mixers. The three mixers deliver the audio IF (Intermediate Frequency) outputs representing the stimulating signal, reflected signals at Channel 0 and the received signal at Channel 1. These three audio signals pass to a TLV320 audio codec that digitises the audio into an I2S data stream that is passed to the STM32F072 microcontroller. It is this microcontroller that performs the maths, drives the display and manages the operation of the NanoVNA.

### NanoVNA-H

The lack of a ready-built model limited the enthusiasm for the original NanoVNA design. However, that changed when Hugen adapted the design for commercial production. The new design was cheaper to produce, had an improved touchscreen, PCB layout and screening. The frequency coverage was also increased to add the 300 – 900MHz band, by utilising harmonics of the clock signal. The new version also supported computer control and the export of touchstone files. This revised version is the one you'll find most widely available using the name NanoVNA-H, and is available with both 2.8in and 4in touchscreens, **Fig.1-10**. There have been a few poor-

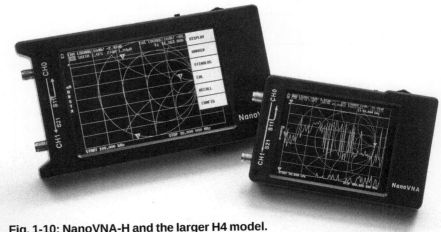

Fig. 1-10: NanoVNA-H and the larger H4 model.

quality clones of this model so you should purchase from a reliable supplier, rather than just seeking out the lowest cost.

## NanoVNA-F

Although using the NanoVNA name, this model is significantly different as it uses a 4.3in IPS screen, different processor running a RTOS (Real Time Operating System) and has a different command structure. This model is available from AliExpress and the firmware can be found here:
https://github.com/flyoob/NanoVNA-F

## NanoVNA V2

The NanoVNA V2 is an entirely new design that HCXQS developed in collaboration with OwOComm. I've shown the block diagram in **Fig.1-11**. An updated signal stimulus and mixer local oscillator chain enable a broader frequency coverage, without reverting to the use of harmonics.

The Version 2 test signal chain begins with an Si5351 programmable clock for frequencies below 300MHz. Above 300MHz the V2 changes to a pair of ADF4350 microwave synthesisers. The three SA612 mixers in Version 1 are replaced with a single AD8342 broadband RF mixer that employs MXD8641 RF switches to route the stimulus, reflected and through signals to the new single mixer. The low IF output from the mixer is amplified and routed to the ADC (Analogue to Digital Converter) that's built into the STM32F103 microcontroller.

As with the previous model, the microcontroller performs the maths, controls the display and manages the general operation of the VNA. The V2

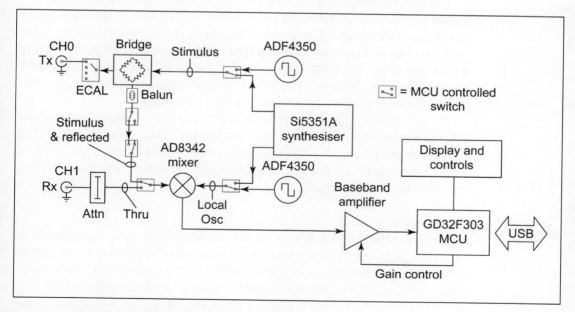

Fig. 1-11: NanoVNA version 2 block diagram.

| Parameter | Specification | Conditions |
|---|---|---|
| Frequency range | 50kHz – 4.4GHz | |
| Frequency resolution | 10kHz | |
| System dynamic range | 90dB (50kHz – 1GHz)<br>80dB (1GHz – 3GHz)<br>70dB (3GHz-4.4GHz) | IF b/w 40Hz, averaging 20x<br>IF b/w 160Hz averaging 5x<br>IF b/w 800Hz averaging 1x |
| S11 noise floor | –50dB<br>–40dB | Below 1.5GHz<br>Above 1.5GHz |
| Sweep rate | 400 points/second<br>200 points/second | 140MHz – 4.4GHz<br>50kHz – 140MHz |
| Sweep points | 10 – 201 adjustable<br>1 – 1024 | Internal<br>Via USB control |
| Battery | 3,200mAh rechargeable | |

Table 1-1: NanoVNA V2 Plus 4 feature summary.

redesign brings several important improvements including faster scans with more data points and a much improved dynamic range on the higher frequency ranges.

I've summarised the important performance features of the V2 Plus 4 in **Table 1-1**. With a top frequency of 4.4GHz, the V2 Plus 4 provides useful coverage of the busy 2.4GHz band. This band is used for Bluetooth, wi-fi, assorted wireless devices as well as the 13cm amateur radio band. Having a VNA that covers the 2.4GHz band can be very useful when testing antennas for any of these devices. However, it's important to remember that your measurements will only be as good as your calibration.

If you intend to use the NanoVNA V2 at 1GHz and above, you should consider investing in a better quality calibration kit. I use the Premium Calibration sets from SDR-KITS (**sdr-kits.net**). Rated for use up to 12GHz, the four-piece set costs under £70 and is supplied with a data sheet containing the important calibration data.

### NanoVNA Version 3

At the time of writing, NanoVNA V3 is in development. This is a completely new design that promises some advanced features, but at a much higher price. Initial estimates are suggesting $400 – $500. However, the new version offers 100kHz to 6GHz coverage and full two-port measurements along with some impressive supporting specifications including 120dB dynamic range to 3GHz.

# NanoVNA Versions

## Summary

The model selection can be simplified to a choice of the current supported models. For those who are mainly working below 300MHz, the cheaper NanoVNA-H or H4 would be excellent choices. The difference between the models is essentially the screen size. Those who need the higher frequency coverage or wider dynamic range should go for the NanoVNA V2 Plus 4, **Fig.1-12**. The larger display and improved dynamic range make this the preferred V2 model. UHF / microwave specialists may prefer to wait for the launch of V3.

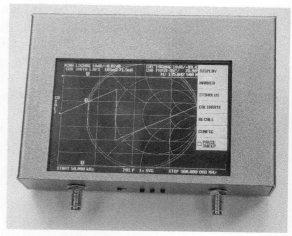

Fig. 1-12: NanoVNA V2 Plus 4.

## NanoVNA Firmware

When dealing with microcontrollers, the firmware is the programming code that tells the microcontroller what to do. Without firmware the microcontroller does nothing! You can think of the firmware as a combination of the operating system and a running program on a conventional PC. The firmware code is stored in non-volatile memory, often on the microcontroller chip. This ensures that the code is ready to run as soon as the microcontroller powers-up.

One of the benefits of this approach is that operation of the NanoVNA or any other microcontroller-based device can be transformed with a change of firmware. However, there is also a danger because any damage to the firmware could cause the microcontroller to be unusable, i.e. it would fail to boot up. This is known as bricking the device. However, the NanoVNA uses the STM32 range of microcontrollers that have their bootloader code factory programmed into ROM (Read Only Memory) during manufacturer. As a result, you can recover from loading incorrect or faulty firmware, simply by entering bootloader mode and uploading the correct firmware. This makes the NanoVNA particularly resilient. At the time of writing, there have been no reported instances of firmware bricked NanoVNAs.

**Identification:** The first step in the update process is to determine your NanoVNA model and current firmware. To do this, start from the main menu and select Config – Version. This will display all the information you need, **Fig.1-13**. The model number is normally shown in larger text at the top of the screen and the firmware version will appear in the main text.

Fig. 1-13: NanoVNA version screen.

### Firmware update – NanoVNA Version 1

Updating the firmware is best done using the free application (DfuSeDemo) from the processor manufacturers, STM. Installing this software provides the USB driver as well as the software application to update the firmware. The software can be found via this link:

https://www.st.com/en/development-tools/stsw-stm32080.html?s_searchtype=keyword

Here are the download and installation steps:

1. Follow the link and select Get Software.
2. This will open a licence agreement page and you may have to give your details.
3. The download will either proceed automatically or you may be sent an email with a download link.
4. When the download completes, unzip the software and run the installer. The next step is to download the latest firmware. At the time of writing, DiSlord was providing the best updates for the popular -H and -H4 models. The releases can be found here:

https://github.com/DiSlord/NanoVNA-D/releases/

The -H and -H4 devices use firmware files in the DFU format (Device Firmware Update). As firmware contributors often change, it's advisable to check with nanovna-users on groups.io to identify the latest version. The DFU update process is very safe and you can use the same process to revert to older firmware versions.

NB: You must choose the correct firmware for your NanoVNA as -H and -H4 are *not* interchangeable.

To update the firmware, you first have to put your NanoVNA into update (DFU) mode. Depending on the model and installed firmware, there are three methods to start the unit in DFU mode as follows:

1. Use the menu to select CONFIG – DFU MODE (**Fig.1-14**); *or*
2. Depress the navigation button whilst powering-up the NanoVNA (**Fig.1-15**); *or*

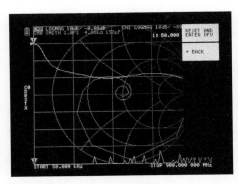

Fig. 1-14: NanoVNA H4 DFU mode selection.

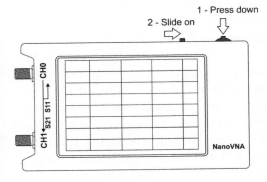

Fig. 1-15: Press the navigation button whilst sliding the power switch on.

# NanoVNA Versions

3. Remove the rear panel, short the BOOT0 and VDD pins, **Fig.1-16**, whilst powering-up the NanoVNA. (NB: Option 3 is the universal solution that will work even if you upload the wrong firmware!)

With the NanoVNA in DFU mode, you should be able to see an entry in Windows Device Manager under USB devices showing: STM Device in DFU Mode, see **Fig.1-17**.

To carry out the update, use the following steps:

1. Open DfuSe Demo, **Fig.1-18**;
2. Check that 'STM Device in DFU Mode' is showing under Available DFU Devices (top left of panel);
3. Move down to the Upgrade or Verify Action section;
4. Click Choose.. and select the firmware DFU file you downloaded;
5. Click the Verify after download checkbox;
6. Click Upgrade to start the upgrade;
7. On completion, quit the program and cycle the power to restart the NanoVNA with the new firmware.

If you encounter any problems repeat the process and make sure you have downloaded the correct firmware for your device.

## Firmware Update – V2

Version 2 of the NanoVNA is a completely new product and you should refer to the official website at:
**https://nanorfe.com/nanovna-v2.html**
for the latest information and updates.

There is also an active user group at:
**https://groups.io/g/NanoVNAV2**
where you can get the latest news and firmware updates.

Whilst V2 uses significantly different hardware, it retains the STM micro-

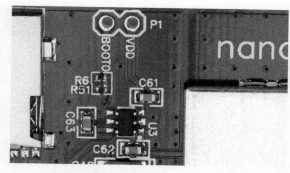

Fig. 1-16: The BOOT0 and VDD reset points on the PCB.

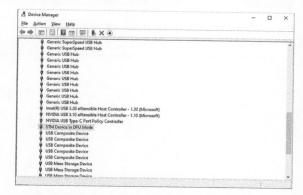

Fig. 1-17: Windows 10 Device Manager showing STM device in DFU mode.

Fig. 1-18: Windows 10 DufSE main screen.

19

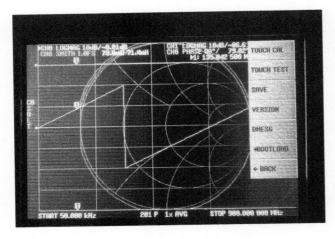

Fig. 1-19: NanoV2 Bootload selection menu.

controller, so you have the safety net of the protected bootloader. Once you have identified and downloaded the latest firmware file, the NanoVNA-QT software provides the simplest upload process. You can download the latest version from the Software section of the NanoVNAV2 site.

Once NanoVNA QT is installed, use the following steps to upload the latest firmware:

1. Start with the NanoVNA disconnected from the PC;
2. Power-up the NanoVNA;
3. From the NanoVNA main menu, select Config followed by +BOOTLOAD or DFU MODE then Reset & Enter or Enter DFU MODE, (**Fig.1-19**);
4. Connect NanoVNA to the PC using a USB cable;
5. Use Windows Device Manager to identify the COM port number;
6. Start NanoVNA-QT;
7. Go to Device menu and select the COM port for your NanoVNA;
8. You will see the message: Device is in DFU mode. Flash new firmware? Click Yes, (**Fig.1-20**);
9. Navigate to the *.bin file you downloaded and click open;
10. You will see the Uploading firmware message.

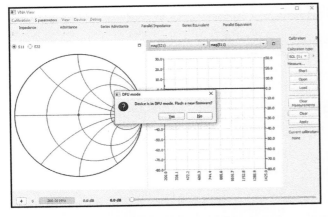

Fig. 1-20: NanoVNA-QT ready to flash firmware.

When you see the 'Done' message you can disconnect and recycle the power to the NanoVNA. If you have any problems, first check that you have the correct firmware, then try uploading again.

## THE SMITH CHART

Philip H Smith first introduced the Smith Chart in an article published in *Electronics* magazine back in 1939. In its original form, the Smith chart enabled RF engineers to calculate complex impedance matching networks using little more than a compass, ruler and pencil. Although modern computing has eased the calculation process, the Smith chart is an extremely useful visualisation tool and remains in regular use by RF engineers.

# The Smith Chart

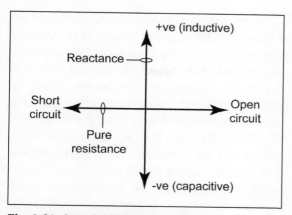

Fig. 1-21: Standard Cartesian chart.

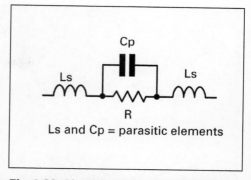

Fig. 1-22: Simplified diagram showing the parasitic capacitance and inductance of a resistor.

## How Does it Work?

The easiest way to get to grips with the Smith chart is to begin by looking at a Cartesian chart, **Fig.1-21**. This chart has a horizontal axis that shows pure resistance with a short circuit on the left and infinity (open circuit) on the right. The vertical axis is used to show reactance with +ve reactance (inductive) moving up from 0 to infinity, and –ve reactance (capacitive) moving down from 0 to –infinity. This type of representation is useful in RF calculations because all networks, even simple resistors, have reactive properties. The resistor leads have inductance and there will be some stray capacitance across the resistor, **Fig.1-22**. Whilst these values may be trivial at low frequencies, they become increasingly significant as the frequency increases.

In **Fig.1-22**, the impedance Z, is a complex value that comprises a resistive element and a reactive element. This would usually be defined by two numbers such as R and j, commonly called the real (R) and imaginary (j) parts. The j in this case represents the reactance with +ve values indicating inductive reactance and –ve values showing capacitive reactance.

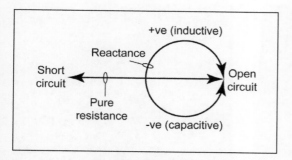

Fig. 1-23: Cartesian chart with the infinity axis joined.

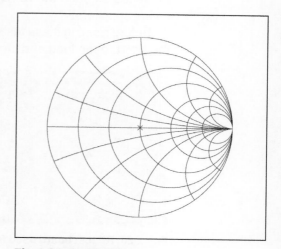

Fig. 1-24: Smith Chart.

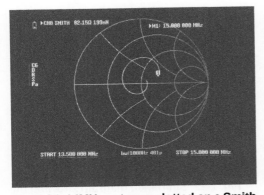

Fig. 1-25: 14MHz antenna plotted on a Smith chart.

To create the Smith chart, Philip Smith realised that the top and bottom of the vertical axis and the right-hand end of the horizontal axis all ended at infinity. His simple idea was to join the infinity points of the chart by changing the vertical axis into a circle that joined the right-hand (infinity) end of the horizontal axis, **Fig.1-23**. With the reactive axis forming a circle, all the graduations on that axis are also circles, hence the familiar Smith chart layout shown in **Fig.1-24**. To make the Smith chart as versatile as possible, the scales are normalised to 1, with the centre of the chart used to represent the characteristic impedance of the network being measured. For most RF work and when using the NanoVNA, the centre dot of the chart represents a pure resistive 50Ω. If you were to plot the response of a pure 50Ω resistance on a Smith chart, you would have a single dot at the centre of the graph. However, real circuits always have some stray reactance (capacitance and inductance) that will pull the trace away from the horizontal resistance line. The reactance is frequency dependant, so will display as a curve on the Smith Chart. I've shown a plot of a 14MHz antenna in **Fig.1-25**.

### Interpreting the Smith Chart

The secret behind the success of the Smith chart lies in its ability to simultaneously display the resistive and reactive values of any network or component.

The easiest way to get started with the Smith chart is to look at the traces produced by single components connected in parallel with CH0. For these tests I used the RF Demo Kit, **Fig.1-26**, that's available from many online suppliers.

Before running these tests you will need to calibrate your NanoVNA using the short, open load and through connections on the test board, **Fig.1-27**.

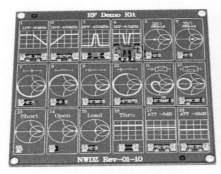

Fig. 1-26: RF test board used for experiments.

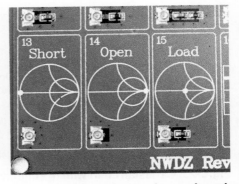

Fig. 1-27: SOLT loads on the test board.

# The Smith Chart

**Capacitor:** For this test, I set the NanoVNA to sweep from 50kHz to 300MHz and displayed the Smith chart results. A helpful way to think about simple components is to consider how they would behave at DC and an infinitely high frequency. A pure capacitor will look like an open circuit at DC because it is simply two closely-spaced plates. However, at an infinitely high frequency, a capacitor will look like a short circuit. Therefore, we would expect the capacitor plot to start at infinity, i.e. the right-hand end of the central horizontal line on the Smith chart. It should also appear as a pure short circuit at a very high frequency and so would be plotted at the other end of the horizontal line. What about intermediate frequencies values, where will they plot?

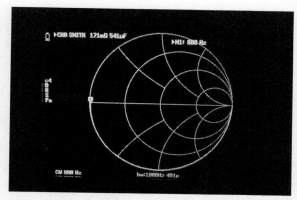

Fig. 1-28: Smith chart showing a 100pF capacitor (50kHz – 300MHz).

In **Fig.1-28**, I've shown a Smith chart plot of a wide sweep from 50kHz to 300MHz connected to a 107pF parallel capacitor from the test board. At the 50kHz end of the sweep the plot starts, as expected, at the right-hand infinity point. As the sweep frequency increases, the trace progresses around the circumference of the chart. This occurs because the capacitor is a predominantly reactive component so the impedance is frequency dependent. If you extend the plot beyond 300MHz, you will find that the plot moves into the top half of the chart and moves away from the circumference. This indicates that the inductive and resistive elements are starting to show themselves.

**Inductor:** Still using the test board, I connected an inductor in parallel with CH0. An inductor has the opposite reactive characteristics to a capacitor. A pure inductor would be a short circuit at DC and an infinitely high impedance at high frequencies. I've shown a plot with the 670nH test board inductor in **Fig.1.29**. Here you can see that the trace starts, as predicted, close to the left-hand, short circuit, end of the chart and progresses close to the circumference. You may be wondering why it doesn't quite reach the edge of the chart. This is due to the DC resistance of the inductor coil. As with the capacitor, you can see that the trace starts to deviate at higher frequencies due to the effects of stray capacitance.

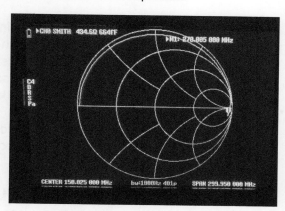

Fig. 1-29: Smith chart showing a 670nH inductor (50kHz – 300MHz).

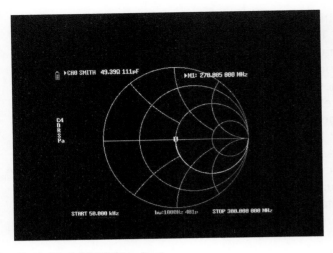

Fig. 1-30: Add R&C to chart.

**Resistor and Capacitor in Series:** In this plot, **Fig.1-30**, The test board has a 50Ω resistor and 100pF capacitor connected in series between the centre pin and ground. As you can see, this trace still follows a circle but, instead of following the outside circumference of the chart it follows the circle that passes through the 50Ω point on the chart. That circle is known as the constant resistance circle.

These simple experiments show you how the Smith chart displays complex impedances, but the next step is to use the Smith chart to solve problems.

### Application

The prime application for the Smith chart is as a design aid for RF matching networks. Early adopters of the Smith chart used a pencil, ruler and compass, but we have access to an excellent, free, software design tool called SimSmith, **Fig.1-31**. As the name implies, this is a circuit simulator packages that uses the Smith chart. SimSmith can import S parameter data from your NanoVNA and combine that data with the circuit simulator to help design matching networks. SimSmith is a free download from:
**http://www.ae6ty.com/Smith_Charts.html**

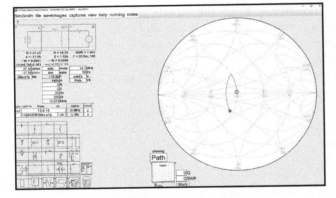

Fig. 1-31: SimSmith software main screen.

The site also includes links to the excellent YouTube tutorials by Larry Benko, W0QE. You should also check the NanoVNA videos by Alan Wolke, W2AEW. These tutorials provide comprehensive coverage of using SimSmith and the NanoVNA.

I've provided a detailed guide for transferring data from your NanoVNA in the Practical section of this book.

## S PARAMETERS SIMPLIFIED

When examining RF devices, engineers commonly use what are called S parameters, or scattering parameters. These have been a source of

# S Parameters

confusion for those new to RF measurement and VNAs in particular. However, the principles are straightforward, as I'll show here.

## Analysing RF Networks

Measuring the performance of RF circuits using scalar values such as simple voltage and current becomes increasingly difficult as the frequency increases. This is due to the reactive nature of connecting cables and devices at RF. Any mismatches in the test cable or the device under test result in standing waves that will compromise simple voltage and current measurements.

An alternative approach is to treat the RF signals as waves. With this approach, we know that a wave travelling into a perfectly-matched load will have no reflections, and the power will successfully transfer to the load. However, any mismatch in the load or connecting cable will result in part of the original signal reflecting (*scattering*) back towards the source, **Fig.1-32**. If the reflected wave arrives in phase with the outgoing wave, the two will combine to increase the apparent value of the outgoing wave.

To take an extreme example, if we were to apply a signal to a short-circuit ¼λ coax line, most of the signal would reflect from the short-circuit and arrive back at the source and in-phase with the source. Therefore, the two waves would add together and double the voltage at that point, making it look like an open circuit. From this example, you can see that we need to measure the amplitude and the phase of the forward and reflected waves to calculate the true impact of an RF circuit.

S parameters are simply a set of numbers that show how the amplitude and phase of the forward and reflected signals vary over a range of frequencies. This data can then be used to calculate and display familiar measurements such as return loss, VSWR, through gain / loss, impedance, etc.

## VNA Measurements

A VNA is the instrument of choice for RF network measurements because it can measure the amplitude and phase of both the forward and reflected waves over a specified frequency range. To achieve this, VNAs use directional couplers to separate the forward and reflected signals, **Fig.1-33**.

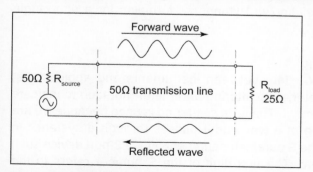

**Fig. 1-32: Reflected wave from a mismatched load.**

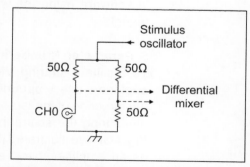

**Fig. 1-33: Resistive directional coupler.**

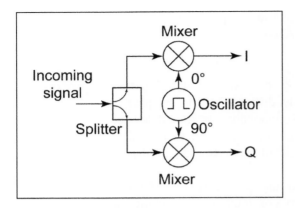

**Fig. 1-34: Quadrature mixer.**

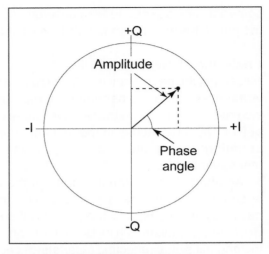

**Fig. 1-35: Polar plot of I and Q values.**

With the signals separated, the next stage is to determine the amplitude and phase of each path. In the NanoVNA, and many other VNAs, this is achieved using the same quadrature mixing that we see in SDR receivers. In some cases this is done with analogue mixers but NanoVNA uses digital quadrature mixing. I've illustrated this in the block diagram shown in **Fig.1-34**. Here you can see that the measured signal feeds two identical mixers that have their local oscillators shifted by 90°. This is called *quadrature mixing*, because one mixer is a quarter of a wave behind the other.

The output from the mixers are the I (In-phase) and Q (Quadrature) signals. By plotting the IQ values using a polar graph, we can illustrate how to convert the IQ data into amplitude and phase values, **Fig.1-35**. In this example, the I signal value is plotted along the horizontal axis, whereas the Q value uses the vertical axis.

If we then use horizontal and vertical dotted lines to join these points, we get a new value, which is the true amplitude of the signal. The phase of that signal is indicated by the angle between the new point and the horizontal axis. The conversion between IQ values and amplitude / phase can be done in software using the well-known Pythagoras formulas for a right-angle triangle.

### S Parameter Numbering

To facilitate sharing VNA data between instruments and applications, engineers use S parameter, or scattering parameter, notation to indicate the direction of measurement. The term *scattering* is used because we are measuring the scattering of a wave as it passes through a system. In **Fig.1-36** I've illustrated the S parameter numbering for a 2-port device such as an amplifier, filter, etc. The first digit of the S number refers to the measurement port, whilst the second digit shows where the *stimulus*, or

# S Parameters

test signal, was applied. So, for a signal passing through this device, the S parameter is S21, i.e. measured at port 2 and test signal applied at port 1. The parameter for the signal reflected from port 1 would be S11, i.e. the signal is applied and measured at port 1. Likewise, the signal applied to and reflected from port 2 would be S22. When measuring antennas we are usually only measuring S11.

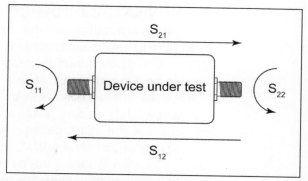

Fig. 1-36: S parameter numbering.

## Normalising

When using Smith charts and other analytic tools it's common practice to normalise the measurements. This is a simple process to change all the measured results so that they are in the same scale. For RF impedance measurements we normalise the results to 50Ω by dividing all the measured impedances by 50. For example, 50Ω would be represented by 1, whilst 100Ω would be 2, and 25Ω would be 0.5. A similar approach is used for amplitude measurements.

## S Parameter File Formats

To simplify the movement of S parameter data between applications, the Touchstone file format has been adopted as the most common standard. The Touchstone format comprises a simple text file where each line holds the S parameter data for a stated frequency. The file format is shown in **Table 1-2**. In addition to the S parameter data, you will usually find an extra line at the beginning of the file that describes the data. Data exported using NanoVNA Saver uses the following format:

\# HZ S RI R 50

\# indicates that the data description follows, HZ indicates that the frequency is expressed in Hz, S confirms that it is S parameter data. RI tells us that the values are pairs comprising real and imaginary values, whilst R 50 is the characteristic impedance of the measurements.

The data on each line is supplied in a strict order, as shown in **Table 1-2**.

Table 1-2: Example of an S parameter Touchstone file.

## S Parameter Summary

I hope you can see that S parameters are quite easy to understand and extremely useful for RF measurements. When combined with appropriate VNA calibration, S parameters can be used to accurately characterise the performance of a wide range of devices.

## NANOVNA CALIBRATION BASICS

Calibration is a routine task for VNA users, but calibration might not be the best term to describe the activity. When we calibrate a VNA we are simply subtracting the effects of the test instrument and its associated test leads. This is similar to the way we zero a multimeter to cancel the resistance of the test leads when taking low-value resistance measurements. However, because VNAs measure phase and magnitude, the process requires more steps and is commonly called calibration.

I'll cover practical NanoVNA calibration in detail later, but the first step in any calibration is to set the frequency range. We then apply three test conditions when prompted. These are: short circuit, open circuit and a 50Ω load. If we intend to do a 2-port measurement, i.e. for an amplifier, filter, etc, we add an isolation and through measurement where the two VNA ports are connected. This type of calibration is commonly known as SOLT (Short, Open, Load, Through) calibration.

Once the calibration is complete, you will have created what's known as a *calibration plane* located at the point where you applied the test conditions, **Fig.1-37**. At this point, the characteristics of the VNA and the test cables are measured and stored. These characteristics are automatically subtracted from any measurements you make, so the VNA will give you a more accurate view of the device being tested.

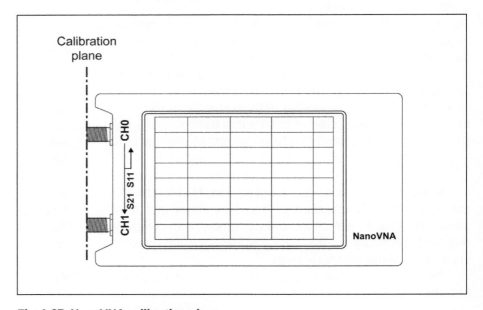

**Fig. 1-37: NanoVNA calibration plane.**

### Supplied Calibration Set

The supplied calibration kit is fine for general use and the loads can be identified as follows:

# NanoVNA Calibration

Fig. 1-38: Open circuit connector.   Fig. 1-39: Short circuit connector.   Fig. 1-40: Load connector.

Open-circuit (**Fig.1-38**) – the connector usually has no centre pin;
Short-circuit (**Fig.1-39**) – when you look at the pin end of the connector, you should be able to see that the pin is bonded to the body;
Load (**Fig.1-40**) – you should be able see insulation around the centre pin and the load often has a taller body. If in doubt, use a multimeter to check for 50Ω resistance between pin and body.

## Measurement Points

One important limitation that you need to keep in mind is the restricted number of measurement points available in the NanoVNA. Depending on which model you have, that could be as low as 101 points. In practice, that means a calibration sweep over the entire 10kHz to 1.5GHz range would result in calibration measurements every 14.8MHz. That's a huge measurement step that could easily miss a whole amateur band or two. For more accurate measurements, you need to set your start and stop frequencies to match the device you're measuring and calibrate using those settings.

## Why are Some Calibration Sets so Expensive?

Those new to VNAs are often horrified at the price of commercial calibration sets. Why are they so expensive and are they really necessary? It all depends on why you're using a VNA in the first place. For example, a design engineer working on the latest mobile phone will need measurement equipment that will make accurate measurements at frequencies in the hundreds of MHz. Likewise, an engineer working on radar collision avoidance systems for vehicles will need accurate measurements at microwave frequencies. Their design work is likely to end up in mass-produced electronics, so it becomes essential that they have precise and repeatable measurement systems.

The accuracy of VNA measurements depends heavily on the quality of the calibration standards you use. Whilst it is relatively easy to build an accurate 50Ω load for the HF bands, it becomes increasingly difficult as the frequency increases. This is because tiny amounts of stray capacitance and inductance can have a significant effect at higher frequencies. I've shown

# NanoVNAs Explained

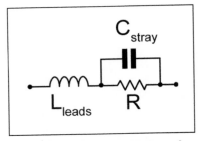

Fig. 1-41: Stray inductance and capacitance of a resistor.

an example in **Fig.1-41**. As it's impossible to build a perfect short, open or load, the high-quality kits employ precision mechanical construction to achieve consistency. This consistent mechanical performance is then supplemented with accurate compensation values that describe precisely how each element of the kit varies from the perfect value. All the professional VNAs are able to accept these compensation values and apply them when carrying out the calibration cycle. The precision construction combined with precise compensation data, drives the high cost of these commercial calibration kits.

Getting back to the NanoVNA, there is little point in using one of these expensive calibration kits. However, SDR-Kits do produce a range of high-quality calibration sets at very reasonable prices (**sdr-kits.net**). These are ideal for upgrading the calibration accuracy of the NanoVNA.

**Connector Savers**

As you become familiar with your NanoVNA you will find it to be an increasingly useful tool. One important precaution you should consider is to add connector savers to each of the NanoVNA SMA sockets. The connector savers comprise an SMA male-female adapter as shown in **Fig.1-42**. By fitting connection savers, the NanoVNA's SMA sockets are protected from general wear and the connection savers can be easily replaced if they show signs of wear.

You should always use a good quality adapter and I recommend the Amphenol (part No: SMA5071A1-3GT50G-50) units as they are a good compromise between price and quality. Avoid using cheap unbranded

Fig. 1-42: Connector savers.

# NanoVNA Calibration

**Fig. 1-43:** NanoVNA PCB mounted SMA sockets.

connectors and cables in your measurement setup as they are likely to lead to inconsistent results and can damage the connectors on your equipment.

Whilst SMA connectors are an excellent choice for operation into the microwave bands, many shacks use UHF, BNC or N-type connectors. If that's the case in your shack, I recommend using a SMA to BNC, UHF or N-type test cables instead of adapters. Whilst you could use adapters on the NanoVNA test ports, the connectors and associated cables with these larger formats will put undue strain on the PCB mounted SMA sockets. See **Fig.1-43**.

## Test Cables

To achieve consistent and accurate results with the NanoVNA (or any VNA), you should use good quality test leads. These are available ready-made from many suppliers, but I'd advise against using unbranded leads. The lower quality leads are likely to give inconsistent results due to connectors that are out of tolerance and cables that change characteristics when flexed. When using SMA connectors, I mainly use leads that employ RG-316 cable. RG-316 is an excellent quality cable and the small diameter makes for flexible leads that are ideal for use with the lightweight NanoVNA, **Fig.1-44**.

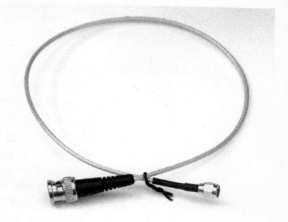

**Fig. 1-44:** Standard SMA to BNC test cable.

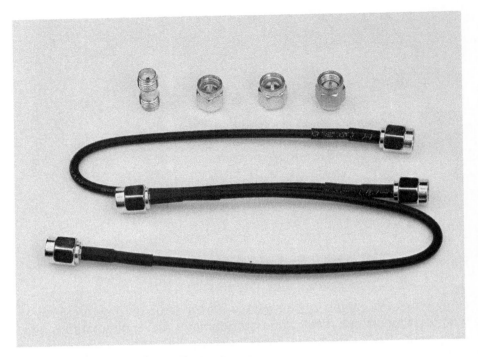

**Fig. 1-45: Supplied calibration kit.**

**Calibration Kits**
Most of the popular NanoVNA versions are supplied with a pair of test leads along with an SMA male calibration kit that comprises a short-circuit, open-circuit, a 50Ω load and a through connection for joining the SMA test leads together. I've shown an illustration of the standard calibration kit in **Fig.1-45** above.

If you want to improve your measurement accuracy at a reasonable price, I recommend the VNA calibration sets produced by SDR-Kits (**www.sdr-kits.net**). The range of calibration kits they supply are excellent quality, reasonably priced, and ideal for use with the NanoVNA.

**Calibration Memories**
As an aid to calibration, the NanoVNA includes five to seven memories, (depending on firmware) that can store the calibration data for different test scenarios. This can be a valuable time saver for common measurement setups. In addition to saving the calibration data, the memories also store the VNA configuration, i.e. the frequency range and display trace settings. This makes the calibration memories a very convenient method of storing and recalling complete test setups.

You should also note that calibration memory 0 loads automatically during power-up, thus making it the ideal place to save a full range calibration for quick measurements. By doing this, you will be able to make coarse measurements anywhere in the NanoVNA's frequency range without

# NanoVNA Calibration

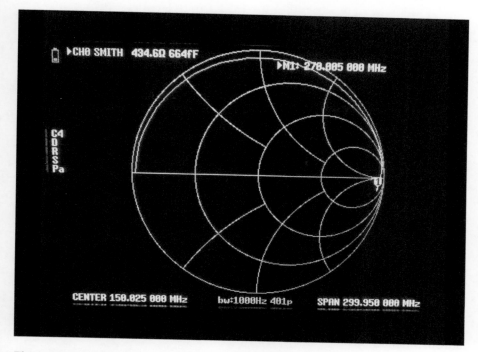

Fig. 1-46: Calibration status is shown on left of the screen.

recalibrating. However, you will need to revert to calibrating a narrower working frequency range for more accurate results.

## Calibration Indicator
The NanoVNA V1 calibration status is indicated by text on the left-hand edge of the display, **Fig.1-46**. For example when using the calibration stored in memory 0 the indicator will show C0. If you change the stimulus frequency range the indicator will change to a lower case c. This indicates that the NanoVNA is still using the original calibration data, but it is being interpolated to match the new frequency range.

The NanoVNA V2 has the same indicator but this is supplemented with the letters SOLT to indicate the type of calibration. SOLT is the full calibration using short, open, load and through. If the calibration didn't include the through test the indicator would show SOL.

## Practical Calibration
Before we start, make sure you have test leads and the calibration kit you plan to use. Let's begin by running a wide-range calibration sweep and storing the results in memory 0. The first step is to configure the STIMULUS to the desired sweep range. For the popular NanoVNA-H4 that would be 50kHz – 1.5GHz, but you should adjust this to match your model. Here's the menu sequence:

# NanoVNAs Explained

1. From the top menu select STIMULUS – START
2. Enter 50k
3. Select STOP
4. Enter 1.5G

Although the display settings don't affect the calibration, they will also be saved in Memory 0, so it's worth setting this up to a useful combination. I suggest setting using TRACE 0 to show SWR and TRACE 1 to show Return Loss. We'll also setup TRACE 2 to show the through loss / gain. Here's the menu sequence:

1. From the top menu select DISPLAY – TRACE – TRACE 0. Make sure it's ticked.
2. Click BACK – FORMAT – SWR
3. Click BACK – CHANNEL – tick CH0 Reflect
4. Go back to main menu and select DISPLAY – TRACE – TRACE 1. Make sure it's ticked.
5. Click BACK – FORMAT LOGMAG
6. Click BACK – CHANNEL – Tick CH0 REFLECT
7. Return to main menu and select DISPLAY – TRACE – TRACE 2. Make sure it's ticked.
8. Click BACK – FORMAT LOGMAG
9. Click BACK – CHANNEL – Tick CH1 THROUGH

That completes the configuration, so we can start calibrating as follows: From the main menu select: CALIBRATE – RESET (**Fig.1-47**). This is an important step and it removes any existing calibration values.

Fig. 1-47: **Calibration reset.**

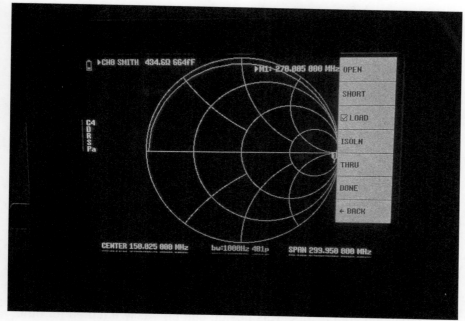

Fig. 1-48: Calibration menu.

1. Next select CALIBRATE and you will see the calibration steps listed, **Fig.1-48**;
2. Connect an open circuit load to CH0 and click OPEN;
3. Wait for the tick to display against OPEN;
4. Connect a short-circuit load to CH0 and click SHORT;

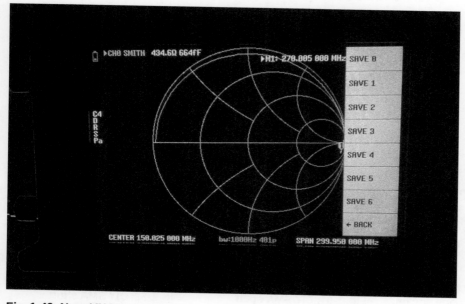

Fig. 1-49: NanoVNA save screen.

5. Wait for the tick to show against SHORT;
6. Connect a 50Ω load to CH0 and click LOAD;
7. Wait for the tick to show against LOAD;
8. Leave the 50Ω load connected to CH0 and connect a second 50Ω load to CH1. Click the ISOLN;
9. Wait for tick to show against ISOLN;
10. Next connect CH0 to CH1 and select THRU;
11. Wait for the tick to show against THRU;
12. Click DONE followed by SAVE 0 (this will save the calibration to memory 0, **Fig.1-49**).

### Homebrew Calibration Kits

It is perfectly feasible to build your own calibration sets, particularly for the HF and VHF bands. One of the simplest techniques is to use flange type panel mounting connectors, **Fig.1-50**. You can generally find these with male and female connectors.

Here's a description of how to build an SMA calibration set based on panel mounting connectors. This design can be easily adapted for other connector systems, but you should ensure that all connectors are from the same manufacturer's range.

**Open Circuit Load (Fig.1-51):** Cut or grind the centre-pin so that it's flush with the insulation. Be very careful to ensure there are no metal filings left on the insulation.

**Short Circuit Load (Fig.1-52):** Cut the centre pin so that it's about 2mm proud of the insulation. Tin the exposed centre pin and the flange area. You may need to apply additional flux to tin the flange. Prepare a strip of coax earth braid or solder wick. Solder the braid to the flange and the centre pin to give a short circuit.

Fig. 1-50: SMA flange socket.

Fig. 1-51: DIY SMA open circuit.

Fig. 1-52: DIY SMA short circuit.

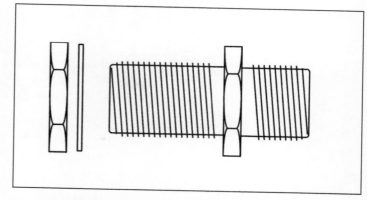

Fig. 1-53: DIY SMA load.

Fig. 1-54: SMA panel mount through connector.

**50Ω Load (Fig.1-53):** Cut the centre pin so it's about 2mm proud of the insulation. Flux and tin the flange. Use either two 100Ω or four 200Ω resistors to create the 50Ω load and solder these between the centre pin and the flange. *NB:* You should select close tolerance metal film resistors and surface-mount size 0805 are ideal and easy to handle. Do not use wire wound resistors.

**Through (Fig.1-54):** For the through connector, use a standard panel mounting through connector.

To complete the project, mount the three loads on to a small metal enclosure. This keeps the kit together making it easier to find, whilst providing protection for the loads.

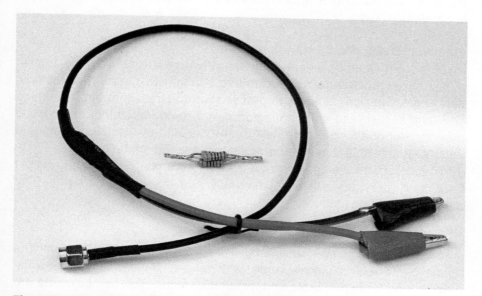

Fig. 1-55: Crocodile clip test lead and load resistors.

# NanoVNAs Explained

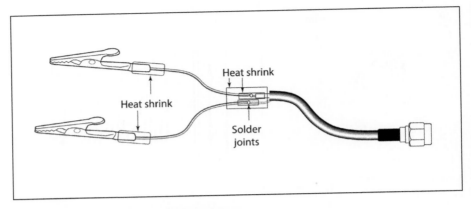

Fig. 1-56: Crocodile clip lead construction.

### Crocodile Clip Test Leads
When making measurements on HF wire antennas or checking components it can be helpful to have a pair of test leads with crocodile clip connectors, **Fig.1-55**. These are very simple to make and I've shown the construction technique I use in **Fig.1-56**. The ideal wire for the crocodile clips is the type used for multimeter test leads. To insulate and protect the joint between the coax and wires, I use plenty of heat-shrink sleeving.

### Calibrating with Crocodile Clip Test Leads
For the short and open calibration, you simply open or short circuit the crocodile clips. However, you will need to make a 50Ω load. This can be as simple as a 50Ω carbon or metal oxide (but not wire-wound) resistor. However, as 50Ω is not an E12 or E24 value, I used three 150Ω 1% resistors in parallel.

### Electrical Delay (Port Extension)
If you have a situation where your NanoVNA is calibrated for the current measurement task but you need to use a longer test lead, the electrical delay parameter offers a way to proceed without recalibrating. Assuming your longer test lead and connectors are a good impedance match, the main impact of a longer cable will be the addition of a small propagation delay or phase shift that's proportional to the length of the cable.

The Electrical Delay parameter, more commonly known as a Port Extension, provides a way to manually add the expected delay. The adjustment can be found in:

DISPLAY – SCALE – ELECTRICAL DELAY

Once selected you will see the familiar on-screen numerical pad with the option to add delay in pico seconds (p) or nano seconds (n). If you know the precise length of your cable and its velocity factor, you can calculate the delay. However, it's often more convenient and accurate to use a trial and error method to determine the exact delay. The simplest way to do this is to configure the display with a single trace that's showing the phase at CH0. Here's a guide to doing that:

From the main menu:

DISPLAY – TRACE –

click on the traces to turn them all off, except for:
TRACE 0 – BACK – FORMAT – PHASE

With the NanoVNA in its calibrated state, less the extension cable and with CH0 open circuit, you should see a horizontal line with the marker indicating 90°. If you don't see this, your calibration is in error.

Next, connect your extension cable and the display will change to a sawtooth or a tilted line. You can now start to return the trace to a 90° horizontal line by adding delays in the Electrical Delay parameter.

For a 1m test cable using RG-316, you can expect the delay to be a few ns. As you get close to the correct figure, the sawtooth will change to a tilted line. If the right of the line is lower than the left you need to add more delay and vice versa if it's above. Once the line is level, you will have found the correct delay and you can start making measurements. If you find you regularly change cables, it may be worth marking each cable with its delay.

## Calibration Status

To the left centre of the display you will see a series of letters and numbers that indicate the calibration status. The NanoVNA V2 uses a different system so I'll cover both here. Here's a breakdown of the NanoVNA-H status indicators:

| | |
|---|---|
| C0 – C6 | Shows which of the seven calibration memories are being used; |
| c0 – c6 | As above, but the lower case indicates that the frequency range has been changed and interpolation is being used for error correction; |
| D | Directivity calibrated; |
| R | Reflection calibrated; |
| S | Source matching calibrated; |
| T | Through transmission calibrated; |

Here's the NanoVNA V2 calibration status:

| | |
|---|---|
| C0 – C6 | Shows which of the seven calibration memories are being used; |
| c0 – c6 | As above, but the lower case indicates that the frequency range has been changed and interpolation is being used for error correction; |
| S | Short circuit calibrated; |
| – | Open circuit calibrated; |
| L | Load calibrated; |
| T | Through transmission calibrated. |

# COMPUTER CONTROL

## Introduction

Connecting your NanoVNA to a PC adds tremendous versatility and provides access to many advanced features. For example, computer control of the frequency sweeps allows almost unlimited measurement steps, thus allowing much more detailed measurements.

Several software packages are available for the NanoVNA, but *VNA Saver* is the most popular, so I will cover that program in detail. Another valuable tool for those with smartphones is the NanoVNA app that's available for Android and iOS devices. This can be particularly helpful when using the NanoVNA away from the workshop, because smartphone screens are usually easier to read in bright light than the NanoVNA.

## NanoVNA Saver
### Installation

Although initially designed as a supporting application for the export of S parameter data files, NanoVNA Saver has grown into a full-blown VNA control application. The software is very well supported and is enjoying continuous development. NanoVNA Saver is freely available for download, both as source code and as binary files for popular systems.

At the time of writing, binaries were available for Windows PCs (32 and 64-bit), Mac and Linux. The latest distributions are stored on GitHub and accessed via the following link:

**https://github.com/NanoVNA-Saver/nanovna-saver/releases**

NanoVNA Saver can also be installed on the Raspberry Pi, as I will show you later.

**Windows users:** The NanoVNA Saver software is entirely self-contained, so there is no installer to run. Choose the x86 file for 32-bit systems and x64 file for 64-bit systems. Download the zip file and expand it to a directory of your choosing. I recommend storing it in a dedicated folder on the C drive, but avoiding Program Files as that may cause permission problems. To run the software, power up and connect the NanoVNA using the USB-C cable and double-click the program file. After a short delay, you should see the VNA Saver main screen as shown in **Fig.1-57**.

**Linux users:** Download and expand the Linux files into a folder of your choice. Before running VNA Saver, you must make the file executable using one of these two methods:

*Method 1:* From File Manager: Right-click on the file in your File Manager and select Properties – Permissions, then use the Execute tab to select either: Only owner or Anyone.

*Method 2:* With the command line: Open a terminal session and cd to the location of the VNA Saver file.

# Computer Control

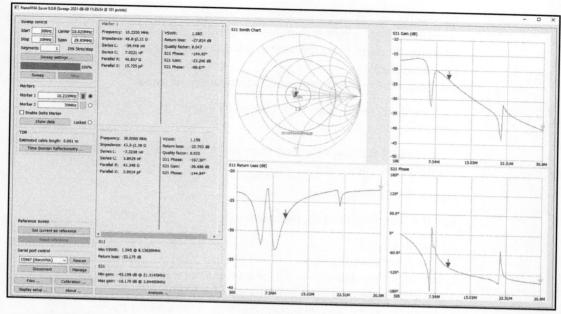

**Fig. 1-57:** NanoVNA Saver main screen.

*Enter:* sudo chmod +x vna-saver

*NB:* Linux users may also need to add their username to the Dialout group or you won't be able to connect to your NanoVNA. This is a simple task as follows:

Open a terminal session and enter: usermod -a -G dialout UserName

*NB:* In the above you must replace UserName with your user name. You will also need to sign-out and in again or reboot the machine for the change of group to take effect. You can check that you are in the Dialout group with this command: getent group dialout

**MacOS users:** Mac users can take advantage of the MacPorts distribution of NanoVNA Saver that's maintained by @ra1nb0w. Here are the steps:
1. If you don't already have MacPorts, that needs to be installed first, see **http://macports.org**
2. Open a shell and enter: sudo port install NanoVNASaver;
3. You can run the program from the shell with NanoVNASaver;
4. The software can also be run from: /Application/MacPorts/NanoVNASaver.app

**Raspberry Pi:** NanoVNA Saver works well on the Raspberry Pi models 3, 4 and 400, but installation requires a few simple steps as shown here. These instructions assume you're using the latest Raspberry Pi OS.

Open a terminal session (Ctl-Alt-T) and enter the following commands sequentially:

41

```
cd ~
sudo apt install -y python3-pyqt5 python3-scipy
git clone https://github.com/NanoVNA-Saver/nanovna-saver
cd nanovna-saver
sudo chmod +x ~/nanovna-saver/nanovna-saver.py
```

You can now run NanoVNA Saver by entering:
`python3 nanovna-saver.py`

To create a desktop link for NanoVNA Saver use the following steps: Create a new file called:

NanoVNA-Saver.desktop

add the following content and save it in the desktop folder:

```
[Desktop Entry]
Name=NanoVNA-Saver
Comment=Runs NanoVNA Saver from the nanovna-saver folder
Icon=/usr/share/pixmaps/openbox.xpm
Exec=python3 /home/pi/nanovna-saver/nanovna-saver.py
Type=Application
Encoding=UTF-8
Terminal=false
Categories=None;
```

### Computer Control Explained

Before going on to utilise the NanoVNA with your PC, it's helpful to have an outline understanding of how computer control works. When connected via the USB port, the computer receives S parameter measurement data from the NanoVNA and sends simple commands to the NanoVNA. The number of measurement points and the total sweep range is determined by the NanoVNA firmware, which is why it's essential to keep this up to date. Note that the NanoVNA must have a valid calibration loaded before attempting to use computer control.

Making measurements with the NanoVNA Saver requires three steps as described here:

**1. Calibration:** You must either load a previous calibration or perform a new calibration using NanoVNA Saver. The calibration is essential to null-out shortcomings in the instrument and the connected test leads. This calibration is in addition to the standalone calibration of the NanoVNA. It is the calibration process that enables the VNA to take accurate measurements.

**2. Sweep Setting:** Adjust the sweep settings to measure the appropriate frequency range. As part of this process, you also set the number of measurement points. The NanoVNA firmware defines the supported measurement points, but NanoVNA Saver can provide a significant increase.

**Computer Control**

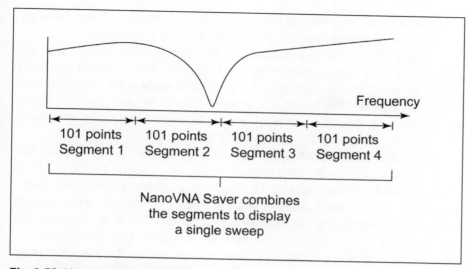

Fig. 1-58: NanoVNA Saver sweep segments.

It does this by cascading several sweeps to appear as one continuous sweep, **Fig.1-58**. The cascaded sweeps are called *segments*, and you have control over the number of segments using the scan controls. The use of scan segments provides an almost limitless number of measurement points at the expense of increased sweep time.

**3. Display Results:** During the sweep, NanoVNA Saver receives a continuous stream of S parameter data from the NanoVNA hardware. Regardless of the selected measurement type, NanoVNA Saver collects the S parameters for both CH0 and CH1. Once the scan is complete, having the full S parameter data available means that you are free to change the measurement type and display format without having to run a new sweep.

**Getting Started**
Before linking the NanoVNA to the computer, you need to complete a full-range calibration of the NanoVNA and store the result in memory 0. This step is required to ensure out-of-range data doesn't get passed to the NanoVNA Saver software.

With the installation complete, it's time to prepare for some measurements. The first step is to ensure that the NanoVNA has connected successfully to your computer. To do this, head to the 'Serial port control' section of the software in the lower-left corner of the main screen. If the NanoVNA has been identified, you will see the port number followed by (NanoVNA), as shown in **Fig.1-59**. If the device is not showing, click the

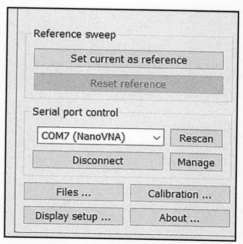

Fig. 1-59: NanoVNA COM port panel.

Rescan button to force a serial port scan. When the port has been identified, click the Connect button to link the NanoVNA to the software.

Whilst in this section, you can use the Manage button to set a few parameters. The most important is the Datapoints which should generally be set to the highest available, so you have increased measurement resolution. However, this decision is a trade-off between resolution and speed; more points will slow the sweep speed. As a starting point, try 201 points with V1 NanoVNAs and 401 for the V2 NanoVNA.

Also in this panel is the bandwidth setting, where I suggest you start with 1000Hz. Lower bandwidths provide greater resolution but at the expense of scan speed. For example, a 101 point scan at 10Hz bandwidth takes nine times longer than the same scan at 1000Hz. One important point to note, when using the Manage panel, is that NanoVNA Saver assumes you have the latest firmware on your NanoVNA. If you have an older firmware, you may find that some options don't work as expected. This often reveals itself as errors during the calibration stage.

### Initial Calibration

When your NanoVNA has successfully connected to your PC, you need to run an initial calibration over the full range of your NanoVNA. As with the standalone calibration, you will need a calibration kit and your preferred test leads. One of the many benefits of NanoVNA Saver is the ability to store a virtually unlimited number of calibration sets. Each of these sets can have notes associated which act as a useful reminder for the future. Those just starting with NanoVNA Saver should use the Calibration Assistant, **Fig.1-60**. This can be found by selecting Calibration and then choosing the Calibration Assistant in the centre-left of the panel. The Calibration Assistant will guide you through the calibration steps and on to saving the results.

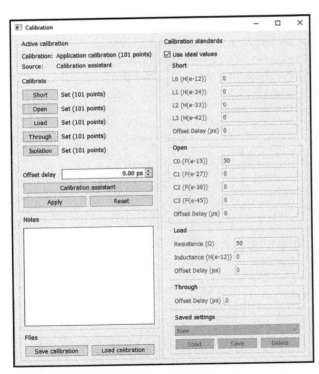

Fig. 1-60: NanoVNA Saver calibration assistant.

### Configuration

NanoVNA Saver's default configuration displays four sets of graphs along with scan data, so can seem a bit overwhelming. I also think a few of the default settings need changing to make the interface clearer, so here's a suggested configuration:

# Computer Control

**Recommended Configuration:** Open Display Settings panel at the bottom-left of main screen, **Fig.1-61**.

Starting from the top left, Options panel, do the following:

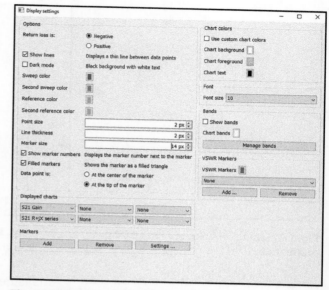

Fig. 1-61: NanoVNA Saver Display panel.

1. *Tick 'Show lines'* – This changes the graph plots from dots to lines;
2. *'Point size' to 2px* – Controls the size of the measurement point dots on the graph;
3. *'Line thickness' to 2px* – Controls the graph line thickness;
4. *'Marker size' to 14px* – Controls the size of the marker triangle;
5. *Tick 'Show marker numbers'* – Displays the marker number above the triangle;
6. *Tick 'Filled markers'* – Makes them easier to see;
7. *Select 'At the tip of the marker'* – moves the marker, so it points at the trace;
8. *Move to upper-right 'Font' and set to 10* – This is the font size used in the entire interface;
9. *Move to bottom-left 'Displayed charts'* and select the charts you want to see displayed.

## Using NanoVNA Saver

The measurement process and NanoVNA test connections remain the same as when using the NanoVNA as a standalone unit. This means you can still use the measurement setups I describe in Part 2 of this book, but NanoVNA Saver gives you far more control of the sweep settings and result processing.

## Sweep Control Panel

NanoVNA Saver provides comprehensive control of the frequency sweep or stimulus used for the measurements. Whilst the measurement points of the NanoVNA are determined by its hardware and firmware version, NanoVNA Saver extends the number of measurement points by employing sweep segments. This breaks the selected sweep range into parts or segments, each of which uses all the measurement point of the connected NanoVNA. The data output from each segment sweep is combined in NanoVNA Saver to produce a more detailed single display.

Let's illustrate the process with an example. In this case, I'm checking the matching of an HF vertical between 3.5MHz and 14MHz. To do this, I need to sweep from 3MHz to 15MHz. If I do this with the standard 101

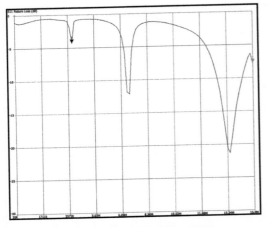

Fig. 1-62: NanoVNA Saver missed 80m resonance.

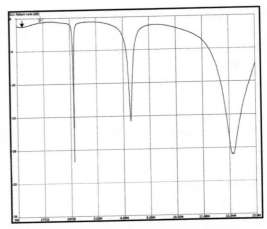

Fig. 1-63: NanoVNA Saver with narrow sweep showing the 80m resonance.

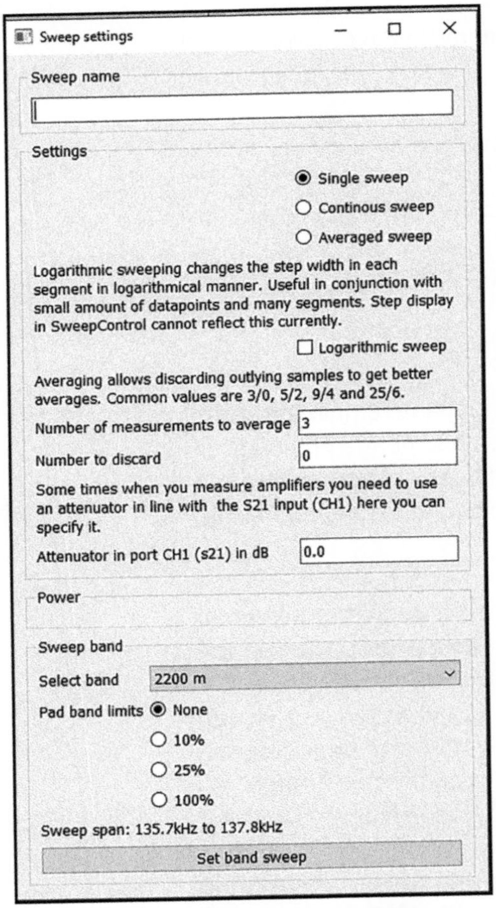

measurement points, I will see a measurement point every 120kHz. That may well be too coarse to see the antenna's very sharp resonance on the 3.5MHz band, **Fig.1-62**. If I use NanoVNA Saver with 10 segments selected, the measurement points increase to 1010 points (101 x 10), so there will be a measurement every 11.9kHz which is much more helpful.

I've shown a screenshot of the resultant sweeps in **Fig.1-63**. As you can see, the segmented sweep clearly shows the 3.58MHz resonance, whereas the single sweep missed it completely!

When setting the sweep range you have the choice of entering the start / stop frequencies or the centre frequency and span. As you enter the desired values, you will see that the measurement step size is automatically calculated and displayed in the Sweep panel.

For greater control of the sweep, click the Sweep Settings button to open the settings panel, **Fig.1-64**. In the first part of this panel you can choose a Single, Continuous or Averaged sweep. The single sweep is the default and is ideal for measuring a fixed devices such as filters, amplifiers, etc. However, when working on an antenna, a continuous sweep is generally more useful. The

Fig. 1-64: NanoVNA Saver sweep settings panel.

next option is the averaged sweep. This lets you average the results from a selected number of scans and can help smooth out noise and spurious results to provide a more accurate interpretation of the device under test.

In addition to a simple average, NanoVNA Saver includes the option to discard outliers. These are often spurious results that are distanced from the true result. By discarding the highest and lowest outliers, the quality of the average result is improved. As a general rule, you should have an odd number of averages and an even number of discards, e.g. 5/2 or 9/4. If you selected 5/2 then NanoVNA Saver would run five scans and for each measurement point, it would discard the highest and lowest results and average the middle three results.

When measuring amplifiers, you will probably need to add an attenuator between CH0 output and the amplifier input to avoid overload. If you note the value of the attenuator and enter it in the Sweep Settings panel, NanoVNA Saver will take that value into account and display the corrected gain of the amplifier, **Fig.1-65**.

Finally, there is an option to select a Sweep Band for one of the amateur radio bands. This is supplemented with a choice to pad the band limits, i.e. to scan a percentage more than just the band edges.

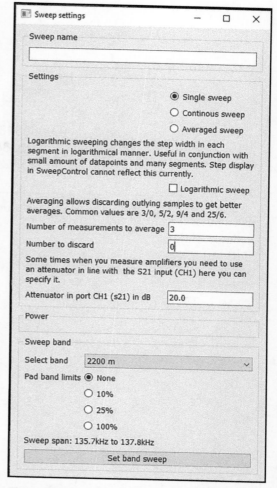

Fig. 1-65: NanoVNA Saver Attenuator entry.

## Data Analysis

This is where the NanoVNA Saver excels, as it uses the larger display and computing power of your PC to process the S parameters from the NanoVNA to produce detailed analysis and graphic displays. The beauty of using the S parameter data is that it can be used to calculate all the measurements we need, i.e., VSWR, return loss, gain / loss, impedance, etc.

The S parameters comprise a list of the measurement frequency plus the real and imaginary values of S11 and S21 for each scan step. In **Fig.1-66** I've shown the first few lines of a single port S11 data file saved using NanoVNA Saver. You can read more in the S parameters section of this book. NanoVNA Saver also adds the facility to save the S parameters in a standard Touchstone format file ready for import into other analysis software such as the excellent SimSmith.

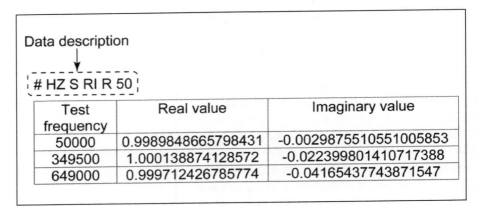

Fig. 1-66: NanoVNA Saver Touchstone file format.

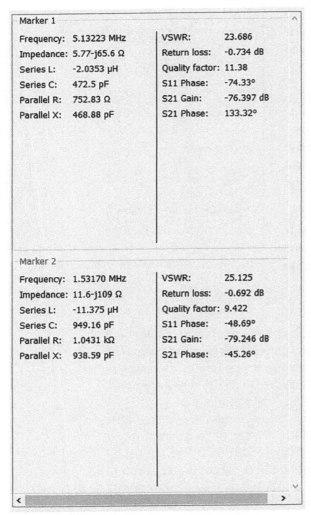

Fig. 1-67: NanoVNA Saver calculated marker data.

**Markers**

Markers provide the facility to take precision, spot, measurements at any point on the scan and are one of the most important tools in NanoVNA Saver. The default configuration uses three markers, but you can add as many as you like via the Marker section of the Display Settings panel. The values displayed by each marker can also be selected using the Marker Settings panel also inside the Display Settings.

The calculated marker values are shown in real-time displayed between the Sweep Control panel and the graphs, **Fig.1-67**. Markers can be dragged across any of the displays using the mouse and, as you move the marker, the associated data in the Marker panel will be updated. To drag a specific marker you first have to select it using the radio buttons in the Markers panel.

In addition to the standard, single point, markers there is provision for a *delta marker*. When activated, this adds a box at the bottom of the Markers data panel that computes the difference between Markers 1 and 2,

## Computer Control

| Delta Marker 2 - Marker 1 | | | |
|---|---|---|---|
| Frequency: | 1.25944 MHz | VSWR: | 0.159 |
| Impedance: | 8.56+j41 Ω | Return loss: | 0.039 dB |
| Series L: | 746.96 nH | Quality factor: | 0.853 |
| Series C: | 732.49 pF | S11 Phase: | -55.62° |
| Parallel R: | 194.9 Ω | S21 Gain: | -1.249 dB |
| Parallel X: | 771.46 nH | S21 Phase: | -2.89° |

Fig. 1-68: NanoVNA Saver delta marker.

**Fig.1-68**. This has many uses from measuring bandwidths through to spotting harmonic relationships.

NanoVNA Saver also includes a dedicated VSWR marker that adds a marker line to the VSWR chart. You can add as many of these as you like and they are particularly useful when using the NanoVNA with antennas or matching networks.

### NanoVNA Saver Analysis Modes

The analysis menu can be found at the bottom of the Markers data panel, **Fig.1-69**. This is an extremely useful extra that adds a few features that are normally only found on expensive VNAs.

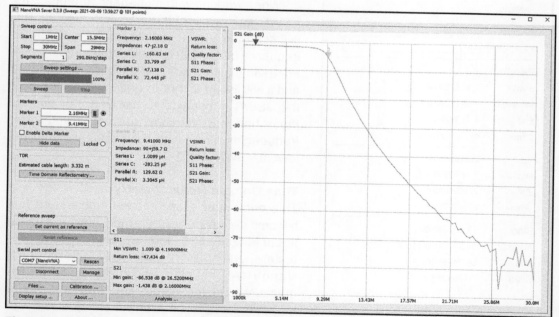

Fig. 1-69: NanoVNA Saver analysis button.

49

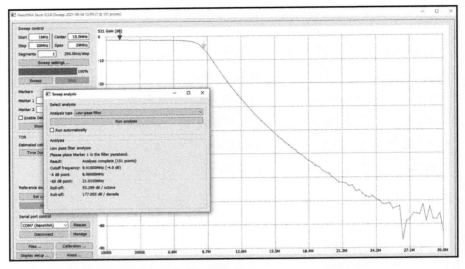

Fig. 1-70: NanoVNA Saver analysis output.

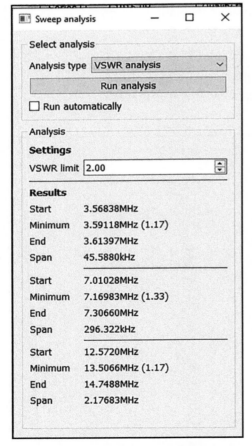

Fig. 1-71: NanoVNA Saver VSWR analysis.

The sweep analysis panel provides an analysis of popular circuit types such as low-pass, high-pass, band-stop and band-pass filters. This is shown in **Fig.1-70**. In addition it can run a peak search to identify and provide measurement values for a peak in any graph.

When working with multi-band antennas, the VSWR analysis tool is particularly useful. This will scan the VSWR results and identify all the frequency bands that fall below your preset value. For example you could set the VSWR limit to 2:1 and it will list up to three frequency bands where the VSWR is better than the limit. For each of those bands it will show the upper and lower frequencies along with the frequency with the lowest VSWR, **Fig.1-71**.

### Time Domain Reflectometry

This once specialised measurement area has become the standard way to measure the distance to cable faults and there are several hand-held testers available for under £50. The barefoot NanoVNA can also be configured for this measurement technique, but NanoVNA Saver makes the analysis that bit simpler and easier to read.

# Mobile App

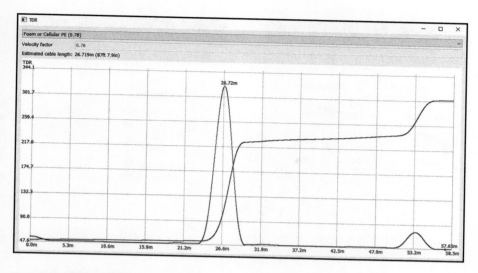

**Fig. 1-72: NanoVNA Saver TDR plot.**

The Time Domain Reflectometry pop-up is selected from the left-hand side of the main panel. When activated it produces a display as shown in **Fig.1-72**. When setting the sweep range you should start at 50kHz and adjust to top frequency to set the maximum length you want to check. To give you an idea of scale, a 900MHz end frequency limits the maximum distance to 5.5m, whereas changing the end to 20MHz increases the range to 250m. I've provided detailed guidance for TDR operation in Part 2 of this book.

### Exporting S Parameters
NanoVNA Saver supports saving the S parameter data in standard Touchstone format so it can be loaded into other data analysis tools such as SimSmith.

### Summary
As I hope you can see from this section, NanoVNA Saver is a significant upgrade for the humble NanoVNA and helps us to make detailed and accurate measurements. The software is enjoying continual development so you can expect to see more features added.

# MOBILE APP
### NanoVNA Web App
This useful web application (app) is available as a free download from the Google Play store for Android devices, or from the author's Github site at:
**https://github.com/cho45/NanoVNA-Web-Client**
for use with PCs.

# NanoVNAs Explained

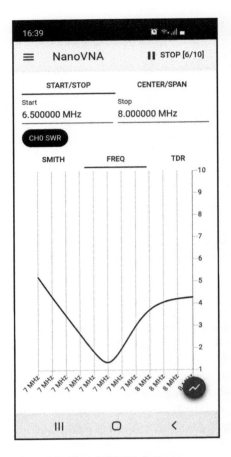

Fig. 1-73: NanoVNA Web App on Android.

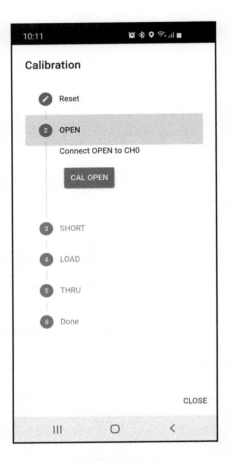

Fig. 1-74: NanoVNA Web App calibration screen.

The app provides an alternative way to control and view data from your NanoVNA, and is particularly easy to use, **Fig.1-73**. I've found the Android version perfect for antenna work outside. Many users find the NanoVNA's screen to be too small or dim for practical use outdoors. However, smartphones generally have very bright and clear displays, so the NanoVNA Web App gives you the best of both worlds, with the NanoVNA making the measurements and your smartphone displaying the results.

Connecting the NanoVNA to your smartphone requires using what's known as an OTG (On The Go) cable. For most modern Smartphones, that is simply a USB-C to USB-C cable. As with the other computer control options, you need to calibrate the NanoVNA via the smartphone app. This is a quick process using the very clearly illustrated steps in the app, **Fig.1-74**. The app also includes five memory slots for storing a selection of frequently used calibration sets.

The app allows the use of multiple traces, and for each, you have the choice of channel (CH0 / CH1) and trace format as follows:

Smith chart
LogMag
Phase
SWR
Linear
Real
Imaginary
R (Resistance)
X (Reactance)
Z (Impedance)

There is also a trace type selection with the following options:

Clear write (This is the default and rewrites the trace for each scan)
Freeze
Max hold
Min hold
Video average
Power average

These are well-chosen options and the video and power averages include an input box where you can specify the number of scans to average.

When making precise measurements, the NanoVNA App has two methods. The first is to use your finger or, preferably, a stylus on the main screen. A crosshair marker and a measurement box appear as soon as you touch the screen. The measurement box showed the results from each trace at the selected position and proved to be a convenient way to get detailed results. In addition, you can define pre-set markers that remain on the display at the chosen frequency, **Fig.1-75**.

I found this particularly useful for antenna tuning work. For example, to optimise my Butternut HF vertical for data modes operation, I could set a marker and then adjust the antenna for best results on the selected frequency. For this to work properly, you may need to narrow the sweep range to make sure there are sufficient sample points to cover your chosen frequency. If you set a marker with insufficient sample points, the app will automatically select the closest sample point. To get a finer spacing of sample points, you can either narrow the sweep range or use the segmented sweep feature.

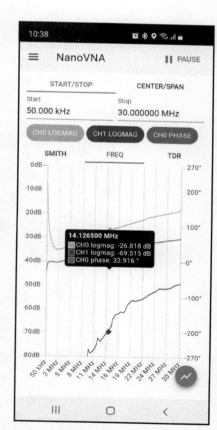

Fig. 1-75: Web App measurement markers.

The segmented sweep is activated via the Resolution menu entry and increases the resolution by breaking the selected sweep range into segments, each of which use the 101 standard sample points. So, setting the number of segments to 5 will result in 505 sample points instead of 101 for the selected frequency range. The five sets of 101 points are automatically cascaded, and the trace is drawn in five steps as a continuous line. This is an excellent way to increase the resolution; the only downside being the increased measurement time.

You can also use the NanoVNA App to download S parameter data to your phone for later transfer to other software for further processing. There are two save options, one for 1-port parameters and the other for 2-port data. The standard Touchstone format is used for the saved files so they are ready to load into programs such as the popular SimSmith.

## NANOVNA AND SIMSMITH

This free software by Edward Harriman Jr, AE6TY, has become the *de facto* standard for designing matching networks for amateur radio.

A tutorial on SimSmith is outside the scope of this book, but I will show you how to export S parameter data from the NanoVNA into SimSmith.

The standard data export format used by most VNA manufactures is based on the format first used by the Touchstone circuit simulator. Whilst the simulator is now obsolete, the file format remains in use. Touchstone files are simple text files, so can be opened and viewed using any text editor. The NanoVNA can save its S parameter data in Touchstone format,

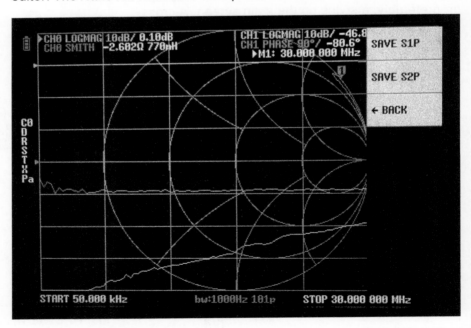

**Fig. 1-76: NanoVNA SD card menu.**

thus making it easy to transfer this data into SimSmith. The technique for saving the data will depend on the NanoVNA model and any control software you may be using.

When using the NanoVNA in standalone mode, the later firmware versions include an option to save the S parameter data to an onboard microSD card, **Fig.1-76**. However, very few NanoVNAs have the SD card slot opened-up in the side of case. It's a simple job to file-out an access slot so you can make use of the SD card. Alternatively, you can use PC-based software such as NanoVNA Saver or the NanoVNA Web App.

In NanoVNA Saver the File Save option can be found at the lower left of the main screen and you should choose the s1p suffix for single port measurements or s2p for 2-port measurements. If using the NanoVNA Web App on Android, you can still save the Touchstone files to your phone using the Save menu. These will normally be saved into the phone's Documents folder.

## Importing Touchstone Files into SimSmith

This is a simple task but it's not obvious for those that are new to the software. SimSmith's simulation is based on the use of circuit blocks that

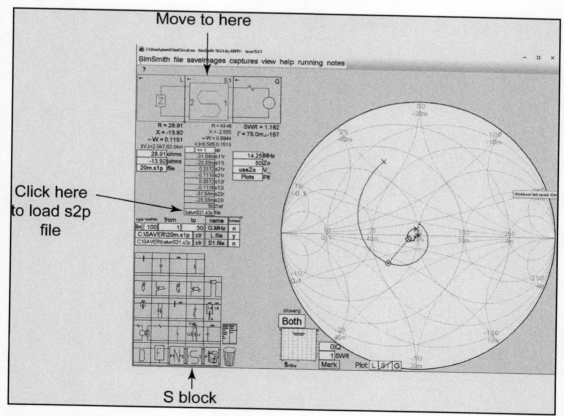

Fig. 1-77: Loading S parameter data into SimSmith.

are dragged and dropped from the pool of blocks to the active area at the top left of the main screen. In order to analyse a circuit that we have measured with the NanoVNA we need to add the appropriate block. As we have captured the circuit's S parameters, it's the S block that we need to drag from the block pool to the simulator section at the top of the screen, **Fig.1-77**.

Once the block is in place, go to the S block parameter table that's located below the block and click on the file box. This will open a file chooser panel where you can browse and upload your S parameter file. Once the file is loaded the S block parameters will be populated and the appropriate plot will appear on the Smith chart.

Once again, SimSmith can be downloaded free of charge from: **http://www.ae6ty.com/Smith_Charts.html**

# Part 2: Practical Measurement Guide

## INTRODUCTION

This part provides concise guidance to help you make the most of your NanoVNA for a wide range of measurement scenarios. This part should be particularly useful for those new to VNAs or those who only use a VNA occasionally.

I have employed a step-by-step approach that covers the VNA settings, calibration requirements and test connections, along with guidance on how to interpret the results.

As with many versatile measurement systems, there's often more than one way to achieve the results, so don't be surprised if you encounter different solutions when reading online information.

My goal is that you will be able to pick up this book and make good use of the NanoVNA and its offshoots without having to do a complete refresher in VNA technology.

### Measurement Tips

I'll offer you a couple of measurement tips that fit well with the use of VNAs:
1: *Do it quick, then do it right.* New users quickly get frustrated with the requirement to calibrate their VNA for every measurement situation. An alternative approach is to store a full-range calibration in memory 0. Using this calibration, you can make rough and ready measurements anywhere within the VNA's frequency range, without recalibrating. This will give you a quick, approximate, result, which may be all you require. If you need to get more accurate, you can refine the result by running a specific calibration and paying attention to the quality of your calibration kit, leads and connectors.
2: *If the results look too good, they probably are!* It is quite common for inexperienced users to get unexpectedly good or bad results. The starting point here is first to understand what results you *should* expect. Think about what you're trying to measure and at least have a mental image of the likely result. With that knowledge, you should be able to recognise an unusual result, so you can go back and double-check your work.

### Markers

Markers are used to make spot measurements at any point on the displayed traces, so they are a vital NanoVNA tool that you need to master. The NanoVNA features six basic markers and a Delta marker that can be used to show the difference between two standard markers. Markers are activated from the main menu using the MARKER option. You then click on the desired marker until the tick appears.

When the markers are active, you will see the measured results displayed at the top of the screen. The easiest way to move the markers is with a touch stylus directly on the display. If you have multiple traces displayed, you should see the markers displayed on each trace. To see the spot measurement for a particular trace, click the marker on that trace and drag it to the desired point. To swap traces, simply click the marker on the desired trace. The system is very intuitive and works well for most situations.

The Delta marker is used to display the difference between any two markers. To use this marker, begin by ticking the DELTA option in the MARKER menu and make sure you have at least two standard markers active. To measure the difference between any two markers, just click on the two markers.

For some measurements, you might be looking for the highest or lowest reading. This can be handled automatically via the MARKERS – SEARCH menu. Here you can search for the maximum or minimum value on a trace. If you then click the TRACKING box, the marker will track the minimum or maximum point. This is ideal to use when you're adjusting a tuned circuit, filter, antenna, etc. The markers also include an option to fix a marker at the START, STOP or CENTRE point on the sweep. This is accessed via MARKERS – OPERATIONS. Finally the MARKER – SMITH VALUE option lets you chose the display format for data from a Smith chart.

## ANTENNA MEASUREMENTS

The NanoVNA is a very powerful tool for antenna analysis. The most obvious use is to check the matching at the antenna but the VNA can do so much more and here are a few examples that I'll cover in this section:

- Measure the antenna matching at the antenna;
- Remotely measure the antenna matching from the shack;
- Optimise ATU settings;
- Measure the feeder loss;
- Measure and cut resonant elements and matching stubs;
- Check for and locate feeder faults.

### Precautions and Preparation

The NanoVNA's light weight make it ideal for use in the field but, if you're working in bright daylight, you might need to make a cardboard hood to

# Antenna Measurements

shade the display so it's easier to read. For best results, colour the inside of the hood matt black. Another handy tip is to support the VNA using a lanyard around your neck; just in case you drop it! The lanyards they hand out at conventions are ideal as they usually include a break link to reduce the risk of injury if you slip and snag the lanyard. An alternative solution for outside work is to link the NanoVNA to your Smartphone with the NanoVNA Web App. This is available from the Google Play Store.

When working with non-earthed antennas, you need to be wary of electrostatic build-up. The NanoVNA inputs have poor ESD protection, so you must discharge the antenna before connecting the NanoVNA. You should also observe the normal precaution of avoiding antenna work when there are storms around.

## Connections

You are likely to need some form of conversion to connect the NanoVNA's SMA socket to your antenna. My preference is to use a conversion test lead, **Fig.2-1**. By that I mean a short test lead that has an SMA plug at one end and the target connector at the other. I favour conversion cables that use RG-316 cable due to the combination of good performance, small size and flexibility. A lower-cost alternative would be RG-174 cable.

If you were to connect the NanoVNA directly to a large diameter cable using UHF (PL-259) or N-type connectors, you would risk damaging the NanoVNA's PCB-mounted SMA sockets. If you prefer to use an adapter, that's your choice, but I recommend using a well-known brand (Amphenol, Cinch, etc), because poor quality adapters will compromise your measurements and can damage the connectors on the VNA.

If measuring LF, MF or HF antennas, a crocodile clip test connection is often the most convenient. I've shown an example in **Fig.2-2** and described the construction and use in the Calibration section in Part 1 of this book (see page 38).

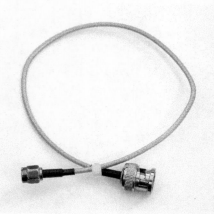

Fig. 2-1: SMA to BNC test lead.

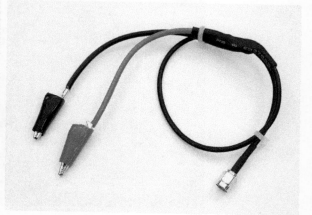

Fig. 2-2: Crocodile clip test lead.

### Settings

When measuring antennas, the most commonly used parameters are VSWR (Voltage Standing Wave Ratio), return loss or a Smith chart. Of these three, the Smith chart is the most revealing, although VSWR or return loss can be easier to use if you are not familiar with the Smith chart.

In most antenna adjustment cases, we're aiming to achieve the lowest VSWR by making physical adjustments to the antenna. In this case, the VSWR or Return loss plots are the quickest to read and give a clear indication of the best match.

Here are the setup steps to configure Trace 0 for VSWR and Trace 1 for Return loss, **Fig.2-3**:

Starting at the top menu:
DISPLAY – TRACE – TRACE 0 – tap until it shows a tick;
Hit BACK – FORMAT – SWR;
Then BACK – CHANNEL – CH0 REFLECT;
Next TRACE – TRACE 1 – tap until it shows a tick;
Hit BACK – FORMAT – LOGMAG – make sure it's ticked;
Click on the main screen to close the menu.

You also need to consider the frequency range used for your measurements. It's important to bear in mind the limited measurement steps available on your NanoVNA. These may be as low as 101 points on a NanoVNA V1. If you set the frequency range too wide, the measurement steps become widely spaced and, in extreme cases, might skip over the

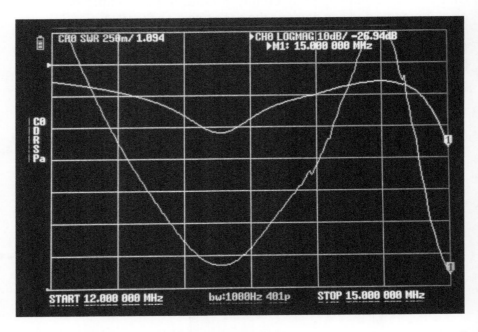

Fig. 2-3: NanoVNA showing SWR and return loss.

# Antenna Measurements

main frequencies of interest. My preference is to set a range that's about 0.5MHz wider than the band you're measuring. For example, when measuring a 14MHz antenna, I would set the NanoVNA to sweep between 13.5 and 15MHz. With the NanoVNA's 101 measurement points, that would provide a measurement every 9.9kHz, which should give a suitably detailed view of the antenna's performance on that band.

If you're working on a multi-band antenna, I suggest you set the sweep for about 0.5MHz above and below the highest and lowest frequencies you want to cover. However, bear in mind that if your antenna has any sharp resonances you may miss them on a wide sweep. In those cases it would be better to split the sweep and use several narrower bands.

Set the frequency sweep starting from the top menu as follows:

STIMULUS – START – Enter frequency
STOP – Enter frequency

## Calibration

When you have your connection method and measurement settings decided, the final step is to calibrate the NanoVNA. Ideally, you should calibrate at the end of your test leads using the appropriate calibration kit. I've shown a selection of home-brew designs in the Calibration section.

As antenna measurement uses just a single port, CH0, you only need to run the Short, Open and Load calibration steps as shown here:

From the top menu:

CALIBRATE – RESET – CALIBRATE
Connect the open circuit load and press OPEN (wait for the tick to appear);
Connect the short circuit load and press SHORT (wait for the tick);
Connect the 50Ω load and press LOAD (wait for the tick) – DONE;
Choose a memory location for the calibration;
Click on the main screen to close the menu.

## Measurement

If you've followed the configuration steps, you should see immediate results when you connect the NanoVNA to the antenna. However, in some cases, the minimum VSWR point may be outside the frequency range of the current sweep. Without clear sight of the current minimum VSWR or resonance, it can be difficult to decide in which direction to tune the antenna. In that case, it's OK to expand the sweep range temporarily to find the current resonance.

Although you may be operating outside the calibrated range of the VNA, the software will use interpolation to calculate acceptable results for this purpose. You can see confirmation of this measurement mode by the lower

Calibration indicator

Fig. 2-4: NanoVNA calibration indicator.

case c on the left-hand edge of the display, **Fig.2-4**. Once you've brought resonance back into the calibrated range of the NanoVNA, you should revert the sweep to the settings you used for calibration. This is conformed by an uppercase C indicator to the left of the display.

Multiband antennas such as the popular Butternut vertical models, exhibit very sharp resonance on the lower frequency bands where the antenna is electrically short. If you use a wideband sweep, the results are likely to be misleading due to the large step sizes. This could result in the sharp resonance being skipped completely. When measuring this type of antenna, you should narrow the sweep range to the band of interest. The NanoVNA will then use interpolation to produce the result, but it is sufficiently accurate to be able to tune the resonance to the desired part of the band. If you want to take a more accurate look at the antenna load on a single band, you can run a fresh calibration over the desired frequency range.

### Remotely Measuring the Antenna from the Shack

There may be occasions when you want to be able to view the antenna matching without having to trek down the garden. The simple solution is to do the Short / Open / Load calibration at the antenna end of the feeder, as shown in **Fig.2-5**.

Although it requires trekking up and down the garden a few times to complete the calibration, it can be worth the effort. By calibrating at the end of the feeder, the feeder becomes part of the measurement system and is

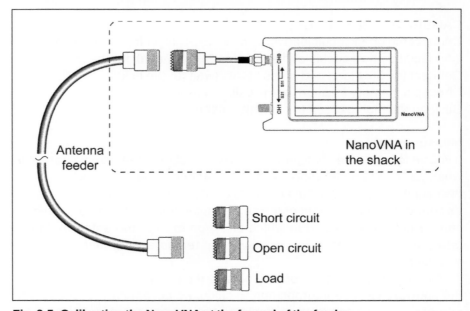

Fig. 2-5: Calibrating the NanoVNA at the far end of the feeder.

# Optimise ATU Settings

eliminated from subsequent measurements. It's the RF equivalent of eliminating the resistance of your multimeter leads by zeroing the meter.

## OPTIMISE ATU SETTINGS

If you use a manual antenna tuning unit (ATU), the NanoVNA is an ideal instrument for determining the optimum ATU settings for each band. It is far better to perform this operation with a VNA than it is to use your rig's RF output and a SWR meter.

This measurement is a case where I recommend using the Smith chart display as this will give a more complete picture of the matching quality so you can see whether to adjust the matching inductance or capacitance.

### Precautions and Preparation

This is a workshop-based measurement, so you can use your NanoVNA stand-alone or by connecting it to the shack computer. Using the shack computer is often to be preferred because the larger display makes the Smith chart much easier to view. The computer software also includes the facility to save the results for future reference.

As with other antenna measurements, you should ensure there's no static charge on the antenna system before connecting the NanoVNA.

### Connections

As mentioned in other configurations, I prefer to use a test lead with an SMA plug for the NanoVNA and the correct plug for the tuner at the other. However, you can also use a good quality adapter at the tuner end of the lead. I've illustrated the connections in **Fig.2-6**.

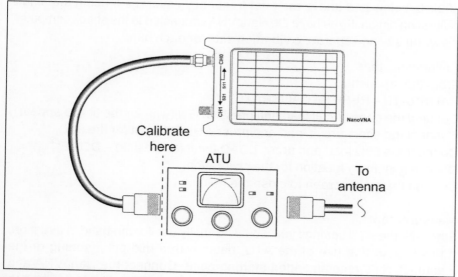

Fig. 2-6: NanoVNA to ATU connection.

### Settings

For this measurement, I recommend using two measurement traces, one showing the VSWR and the other showing a Smith chart. These can be individually activated via the Trace menu. I find that the VSWR trace is ideal for making the coarse measurement to bring the controls close to an ideal match. You can then switch to the Smith chart for the fine adjustment.

Here are the settings:

1: Set frequency range:
Start at the top menu:
STIMULUS – START – ENTER FREQUENCY – STOP – ENTER FREQUENCY.

2: Configure the display for VSWR and Smith chart:
Start at top menu:
DISPLAY – TRACE – click until TRACE 0 is ticked;
BACK – FORMAT – SWR – BACK – CHANNEL – CH0 REFLECT – BACK –
DISPLAY – TRACE – click until TRACE 1 is ticked;
BACK – FORMAT – SMITH – BACK – CHANNEL – CH0 REFLECT.

### Calibration

If you want the best accuracy, you will need to calibrate the NanoVNA for each band. Alternatively, you could perform a single calibration over the entire range of the ATU and rely on the NanoVNA's interpolation for individual band measurements. In most cases this latter approach will give a satisfactory result. As we're using a single port measurement on CH0, there is no need to do the 'Through' calibration step. I suggest you select a calibration frequency range that extends 500kHz either side of each band edge or 500kHz below and above the total frequency coverage if using the single calibration option. If you have the NanoVNA connected to the shack computer, it's worth saving separate calibration sets for each band.

Calibration steps:
From the top menu:
CALIBRATE – RESET – CALIBRATE;
Connect the open circuit load and press OPEN (wait for the tick to appear);
Connect the short circuit load and press SHORT (wait for the tick);
Connect the 50Ω load and press LOAD (wait for the tick) – DONE;
Choose a memory location for the calibration;
Click on the main screen to close the menu.

### Measurement

Start with the ATU settings you would normally use for the band in question. If this is your first use of the ATU, begin with a mid-point setting or the manufacturer's recommended starting point. Connect the NanoVNA and adjust the controls to obtain the lowest VSWR reading. As you approach the optimum setting, activate the Smith chart and continue using the ATU controls to move the trace as close as possible to the centre of the chart. The ATU controls are likely to interact so you will need to experiment to find the best match. Once you've refined the measurement, make a note of the ATU settings and move on to the next band to repeat the process for each band.

## MEASURE ANTENNA FEEDER LOSS

It's good practice to check your feeder integrity regularly and one way to do this is to check the insertion loss. This could be done by connecting an accurate RF signal source to the feeder and using terminating dB meter to measure the level at the distant end, **Fig.2-7**.

However, that type of measurement kit is unlikely to be available to most operators, so we can use the NanoVNA to make an accurate measurement. This technique, requires the use of a spare length of 50Ω coax that's long enough to reach from the shack to the far end of your feeder cable (that's assuming your feeder is also 50Ω!) The cable type is not critical, and RG-58 is a good candidate as it's cheap and readily available. The technique we'll use here is an extreme example of calibrating-out the test lead.

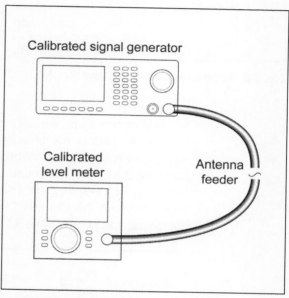

Fig. 2-7: Feeder loss measurement.

### Connections

As with all antenna work, make sure you discharge any static before connecting the NanoVNA to the antenna. You will need either a suitable test lead or adapters to connect the NanoVNA's SMA connectors to the feeder.

### Settings

Our main priority in this test is to measure the insertion loss of the cable, i.e. the loss between CH0 and CH1 of the NanoVNA, but the calibration method we will use will effectively move the calibration plane of CH1 to the end of our test cable, **Fig.2-8**. In addition to measuring the insertion loss, it's also worth measuring the return loss at CH0, as that will give confirm that we a have a good quality impedance match. You should set the STIMULUS range to match the bands for which the feeder is used.

Here are the NanoVNA settings:

# NanoVNAs Explained

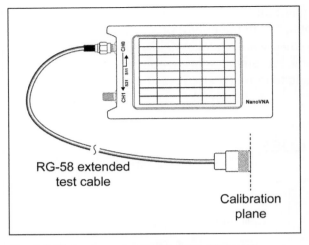

Fig. 2-8: Using an extended test lead for CH0.

Starting from the main menu:
STIMULUS – START – Enter the lowest frequency you want to test;
STOP – Enter the highest frequency you want to test;
BACK – DISPLAY – TRACE – click TRACE 0 until the tick appears;
BACK – FORMAT – LOGMAG – BACK – CHANNEL – CH1 THROUGH;
BACK – TRACE – Click TRACE 1 until the tick appears;
BACK – FORMAT – LOGMAG – BACK – CHANNEL – CH0 REFLECT;
Click on the main display to close the menu.

## Calibration

In this measurement, we will include the extended coax cable as part of the NanoVNA. This is easily achieved in the calibration process. Begin, as always, by resetting the calibration and starting a SOLT test sequence and applying the Short, Open and Load devices to the end of the test cable as shown in **Fig.2-9**. For the Through calibration, CH0 and CH1 are joined together using the extended coax cable, Fig.2-9. When the calibration is complete, you should save it to one of the five internal memories.

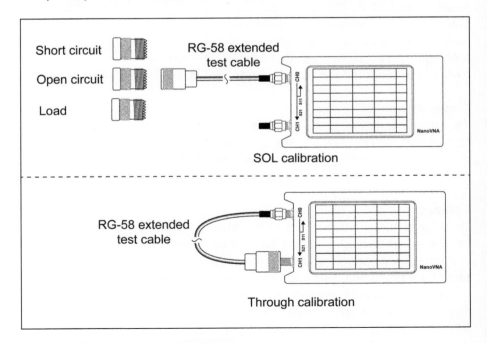

Fig. 2-9: Calibrating the extended test lead.

Here are the calibration steps:

From the top menu:
CALIBRATE – RESET – CALIBRATE;
Connect the open circuit load and press OPEN – wait for the tick;
Connect the short circuit load and press SHORT – wait for the tick;
Connect the 50Ω load and press LOAD – wait for the tick;
With the load still connected press ISOLATION – wait for the tick (NB: Not all firmware has this step);
Connect CH0 to CH1 – press THROUGH – wait for the tick – DONE;
Choose a memory location (SAVE 0 – SAVE 5) for the calibration;
Click on the main screen to close the menu.

**Measurement**

As a confidence check that the calibration has been completed correctly, connect the extension coax cable between CH0 and CH1. If all is well, TRACE 0 should show a loss of very close to 0dB. If it doesn't you should carefully repeat the calibration as you may have made a mistake.

For the measurements, connect CH0 to the shack end of the feeder cable and CH1 to extension coax cable. At the distant end, connect the extension coax cable to the feeder being measured.

The NanoVNA will now display the feeder's return loss and insertion loss. To check whether or not the result is OK you should compare the result with the feeder cable manufacturer's loss tables.

# MEASURING AND CUTTING ¼λ STUBS

Resonant stubs have many uses in antenna designs and the NanoVNA can help you make very accurate cuts. One of the benefits of this technique is that you don't have to rely on generalised velocity factors because you will be directly measuring the resonance of the stub. This is also a useful system for cutting resonant lengths of unknown cable.

The popular ¼λ stub provides an impedance reversal across its length. Therefore a short-circuit at one end of the ¼λ stub looks like an open-circuit at the other, and *vice versa*. I've illustrated this in **Fig.2-10**.

We can measure this transformation in several ways. A popular choice is to use a Smith chart display. In this case, an open or short circuit stub with a wide sweep range, will produce a circle close to the outer edge of the Smith chart. We can then move the marker along the trace with the jog-wheel, to measure the frequency where the trace passes the horizontal axis.

For an open-circuit stub, you need to measure where the trace crosses the left-hand end of the chart's horizontal axis (low resistance), **Fig.2-11**, whereas for short-circuit stubs, it's the right-hand end of the horizontal axis. The Smith chart method is easy to use but the response at resonance is fairly broad.

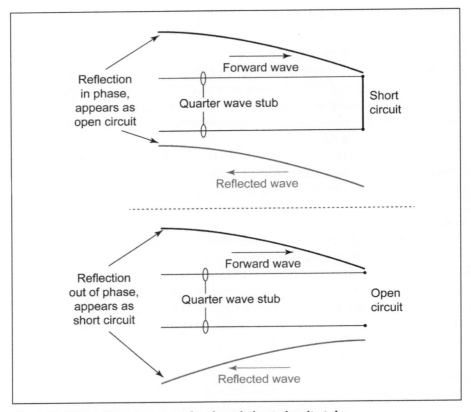

Fig. 2-10: Reflections for open circuit and short circuit stubs.

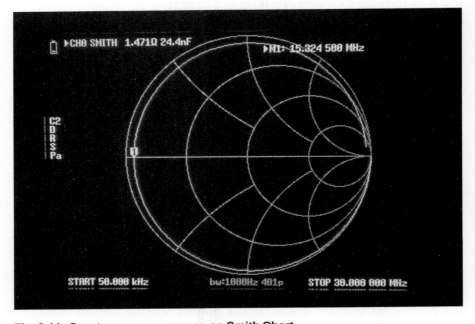

Fig. 2-11: Quarter wave resonance on Smith Chart.

# Measuring 1/4λ Stubs

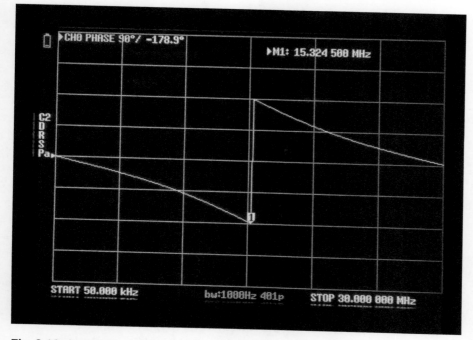

Fig. 2-12: Quarter-wave resonance using phase measurement.

For a more precise indication, I suggest using the phase change method. In this method we use the NanoVNA to measure the phase of the reflected signal on the stub. As the sweep frequency approaches a ¼-wave point the reflected phase approaches 180° and rapidly reverses phase at the resonant length. I've shown a typical sawtooth VNA trace in **Fig.2-12**. As you can see, the trace begins at 0° and moves linearly to −180° where it does a rapid reversal to +180°. This process repeats for each multiple of a ¼-wavelength.

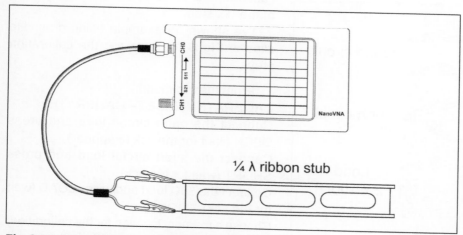

Fig. 2-13: Crocodile clip connection to quarter-wave ribbon.

### Connections

As we're trying to measure a very sharp, length dependent, resonance, it's particularly important to ensure the entire test lead is used when carrying out the calibration. Ideally, we want to transfer the calibration plane to the beginning of the ¼λ stub. One way to do this would be to fit a coax connector to the measurement end of the stub coax.

However, for many HF applications you may need to cut a ¼λ wire or twin feeder stub. In that case, it may be easier to use a test lead with crocodile clips for attaching to the antenna, **Fig.2-13**. Providing you follow the appropriate calibration steps you will still get good results.

### Settings

We will use two traces for this measurement, with TRACE 0 set to display a Smith chart, whilst TRACE 1 will show the phase of the reflection on CH0. When selecting the start and stop frequencies, I suggest you start with a sweep that's 500kHz wider than the desired resonance. I've shown the menu steps to configure the NanoVNA here.

Starting from the top menu:
STIMULUS – START – Enter the lowest frequency you want to test;
STOP – Enter the highest frequency you want to test;
BACK – DISPLAY – TRACE – click TRACE 0 until the tick appears;
BACK – FORMAT – SMITH – BACK – CHANNEL – CH0 REFLECT;
BACK – TRACE – Click TRACE 1 until the tick appears;
BACK – FORMAT – PHASE – BACK – CHANNEL – CH0 REFLECT;
Click on the main display to close the menu.

### Calibration

As we are measuring the characteristics of the reflected signal on CH0, we only need to SOL calibration. However, it is important to perform the calibration as close as possible to the start of the ¼λ stub.

I've shown an example using crocodile clips in **Fig.2-14**. Here are the calibration steps:

From the top menu:
CALIBRATE – RESET – CALIBRATE;
Connect the open circuit load and press OPEN (wait for the tick to appear);
Connect the short circuit load and press SHORT (wait for the tick);
Connect the 50Ω load and press LOAD (wait for the tick) – DONE;
Choose a memory location for the calibration;
Click on the main screen to close the menu.

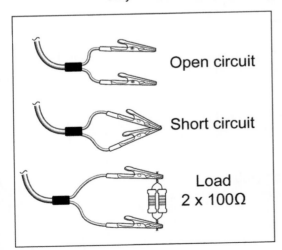

Fig. 2-14: Crocodile clip calibration.

# TDR – Time Domain Reflectometry

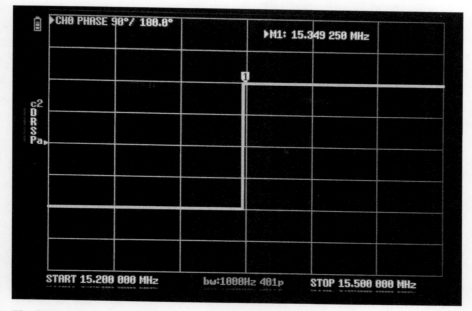

Fig. 2-15: Close view of the phase change at resonance.

## Measurement

To keep the display simple, begin with just the Smith chart (Trace 0) on view. When measuring an open-circuit stub, you should see an arc on the Smith chart's left-hand edge. It's essential that the arc crosses the central horizontal line. If you are measuring a short-circuit stub, the arc on the Smith chart will appear on the right-hand side.

The next step is to turn to the phase measurement and look for the sharp positive to negative spike. This precise ¼-wave length is the frequency where the near vertical trace crosses zero, **Fig.2-15**.

You can use the scroll wheel to move the marker and measure the frequency of the changeover point. The next step is to trim the stub length to achieve the desired resonance. To display a clear spike you will probably need to progressively reduce the STIMULUS sweep width.

## TDR – TIME DOMAIN REFLECTOMETRY
### Measuring the Length of a Cable / Distance to a Fault
### Introduction

This measurement technique is based on the principle that any impedance changes in a cable length will result in a portion of the test signal being reflected back to the source. The impedance change could be due to the open end of a severed cable or some physical damage to a cable.

Providing we know the cable's Velocity Factor, we can calculate the distance to the fault or end of the cable by measuring the time taken for the reflection to get back to the NanoVNA.

This technique is known as Time Domain Reflectometry (TDR) and is used extensively for cable fault location.

### Connection
As with previous examples, begin with a suitable test lead for the cable you are testing. This measurement technique only requires the use of CH0, so a single test lead is required.

### Settings
For this measurement we use a single trace on CH0 and employ the low-pass impulse transform to provide the distance measurement.

For accurate measurements, you also need to enter the velocity factor of the cable. This is necessary because the propagation speed of RF signals through coax is slower than in free space and determined by the construction of the cable. Velocity factors are published for all the popular coax types and are usually expressed as a fraction, i.e. 0.87. However, the NanoVNA requires the velocity factor as a percentage so 0.87 would translate to 87%.

For TDM measurements the STIMULUS frequency span is critical and is set with regard to the expected cable length. The START frequency should be set to 50kHz but Alan Walke, W2AEW, has shared a formula that gives useful guidance on STOP frequency selection:

*STOP frequency = (5850/maximum distance (m)) x Velocity Factor (fraction)*

In **Table 2-1**, I have shown rounded values based on the W2AEW formula and using common velocity factors of 0.85 and 0.66.

Having determined the STOP frequency and Velocity Factor you can use the following steps to configure your NanoVNA.

| Max. cable length | Stimulus STOP frequency | |
|---|---|---|
| | Velocity Factor 0.85 | Velocity Factor 0.66 |
| 5m | 1GHz | 800MHz |
| 10m | 500MHz | 400MHz |
| 20m | 250MHz | 200MHz |
| 30m | 170MHz | 130MHz |
| 50m | 100MHz | 80MHz |
| 75m | 70MHz | 52MHz |
| 100m | 50MHz | 40MHz |
| 200m | 25MHz | 20MHz |

Table 2-1: Suggested STOP values for the common coaxial cable velocity factors of 0.85 and 0.66.

# TDR – Time Domain Reflectometry

Starting from the top menu:
STIMULUS – START – Enter 50k – STOP – Enter chosen STOP frequency;
BACK – DISPLAY – TRACE – Remove all but TRACE 0 by clicking on each trace in turn;
BACK – FORMAT – MORE – LINEAR;
BACK – BACK -TRANSFORM – LOW PASS IMPULSE – TRANSFORM ON (indicated with a tick);
VELOCITY FACTOR – Enter velocity factor as percentage, not fraction, i.e. for 0.87 velocity factor enter 87;
Click on the main screen to close the menu.

## Calibration

For the most accurate results you should run a Short, Open, Load calibration before starting the measurement. I've shown the calibration steps here.
From the top menu:
CALIBRATE – RESET – CALIBRATE;
Connect the open circuit load and press OPEN (wait for the tick to appear);
Connect the short circuit load and press SHORT (wait for the tick);
Connect the 50Ω load and press LOAD (wait for the tick) – DONE;
Choose a memory location for the calibration;
Click on the main screen to close the menu.

## Measurement

Providing you've completed the configuration and calibration steps, you should see an immediate result when you connect a cable to the NanoVNA. In the case of an Open or Closed circuit cable end, you will see a well-defined peak, as shown in **Fig.2-16**. You can then scroll the marker to the

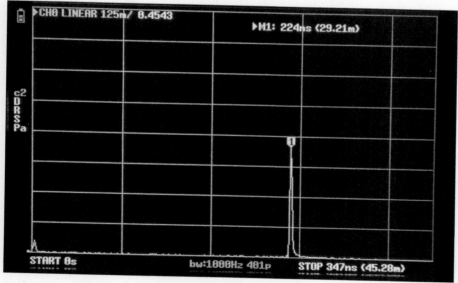

Fig. 2-16: TDR reflected pulse from mismatch.

top of the peak using the thumb-wheel. At that point, the marker display will show the propagation delay and distance to the incident at the top of the screen.

You can also use the marker search facility to lock on to the peak automatically. To use this, go to the top menu and select:

MARKER – SEARCH – MAXIMUM

If you're using the NanoVNA for fault location, you may well encounter multiple peaks or even a soft hump. This typically happens if the cable has suffered moisture ingress. In that case, the distance shown by the left-hand edge of the hump indicates the start of the damaged cable section.

## MEASURING RF SWITCH AND RELAY PERFORMANCE

### Introduction

Amateur radio stations often use RF switches and relays to switch antennas, for transmit / receive switching or filter selection. These switches are often sourced from online sales sites or rallies, so it would be wise to check the performance before installation, **Fig.2.17**.

There are also cases where miniature, non-RF specific relays can be pressed into service. To properly understand the performance of a switch or relay we need to measure the following parameters:

**Forward loss:** This will show us how much signal we might lose as it passes through the switch and it can be a useful check of the contact quality.

**Return loss:** This will show if there are any impedance matching problems in the construction of the switch. This is a particularly useful test when using switches that might not have been designed for RF operation.

**Isolation:** This will show how much signal could leak between the selected port and any other ports. This is particularly critical for high power transmit / receive switching where leakage from the transmitter could damage a sensitive receiver.

### Connections

You will need two test leads with the appropriate connectors to mate with the switch or relay under test. For accurate measurements, it's also

**Fig. 2-17: Coaxial RF switch.**

# Measuring RF Switch / Relay Performance

Fig. 2-18: Terminating plugs.

essential to provide a 50Ω termination for any unused ports on the switch. As a result, you may need a few terminating plugs. These are readily available from the major component suppliers, often for less than £10 each, **Fig.2-18**. You can also build terminating plugs using the designs shown for 50Ω loads in the calibration section.

## Settings

For these tests you'll need to first decide the frequency range you want to measure. You have a couple of options. The first is to select the frequency range of the bands you'll be using. Alternatively, if you have the original manufacturer's data sheet, you could use those values. You'll be measuring two parameters, the first is the through loss and the second is the matching quality as return loss.

Here are the menu steps to configure the NanoVNA to measure the forward loss and the return loss at the input:

From the top menu:
STIMULUS – START – Enter desired start frequency – STOP – Enter desired stop frequency;
DISPLAY – TRACE – TRACE 0 (click till ticked);
BACK – FORMAT – LOGMAG;
BACK – CHANNEL – CH1 THROUGH;
BACK – BACK – DISPLAY – TRACE – TRACE 1 (click till ticked);
BACK – FORMAT – LOGMAG – CHANNEL – CH0 REFLECT.

## Calibration

The calibration requires the full Short/Open/Load/Through calibration between CH0 and CH1. Here are the calibration steps.

From the top menu:
CALIBRATE – RESET – CALIBRATE;
Connect the open circuit load and press OPEN – wait for the tick;
Connect the short circuit load and press SHORT – wait for the tick;
Connect the 50Ω load and press LOAD – wait for the tick;
With the load still connected press ISOLATION – wait for the tick (*NB:* Not all firmware has this step);
Connect CH0 to CH1 – press THROUGH – wait for the tick – DONE;
Choose a memory location (SAVE 0 – SAVE 5) for the calibration;
Click on the main screen to close the menu.

## Measurement

As mentioned in the introduction, to properly check the switch or relay's performance, we will need to take a series of measurements. The first is the forward or insertion loss of the switch, from the common port to the switched ports, as shown in **Fig.2-19**.

For this measurement, CH0 of the NanoVNA is connected to the common port of the switch and CH1 to the first selected port. You should install terminating plugs on all the unused switch / relay ports.

To check for dirty switches, it's worth tapping the switch or relay during measurement to see if that causes any spikes on the NanoVNA display. If it does, you probably have dirty contacts or a worn-out switch with weaked contact tension.

When you've completed the first port of the switch / relay, activate the next port and repeat the measurement, but to the new port. Repeat this

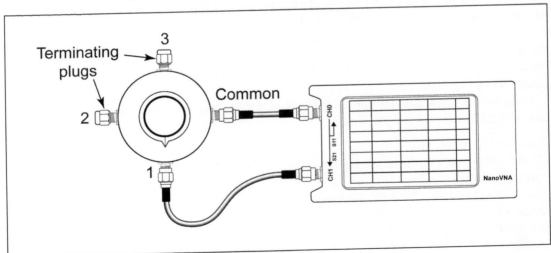

Fig. 2-19: Measuring the forward loss to port 1.

# Measuring RF Switch / Relay Performance

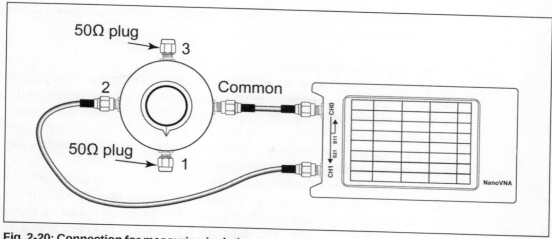

Fig. 2-20: Connection for measuring isolation.

measurement process for each port of the switch / relay. When complete, you will have measured the forward loss to each port of the switch or relay.

Now we can turn our attention to the *isolation*. This is the amount of signal that leaks between the selected port and any other port. For this measurement, we connect the NanoVNA between the common port and an unselected output port, **Fig.2-20**. As with the previous switch measurements, all unused ports should be fitted with terminating plugs.

However, the limited dynamic range of the NanoVNA may well give an unclear result due to its relatively high noise floor, **Fig.2-21**. This could be as low as 40dB on the highest frequency range. For HF to VHF

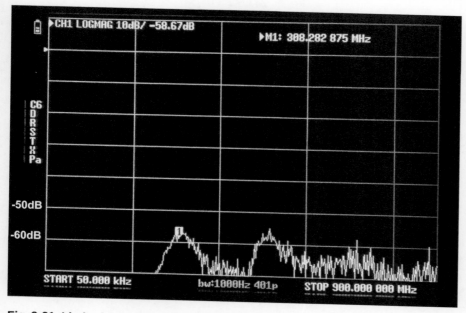

Fig. 2-21: Limited dynamic range of NanoVNA at high frequencies.

77

measurements the dynamic range is much better and a NanoVNA V2 can give a usable dynamic range of 80dB.

### Increasing Dynamic Range

For measurements below 300MHz, it is possible to increase the dynamic range for through measurements by about 20dB. As we can't improve the sensitivity of NanoVNA, the simplest way to effectively increase the dynamic range is to amplify the output of the NanoVNA so that we can apply a higher level test signal to the switch or device we're testing. This requires the use of an external amplifier and an attenuator.

The amplifier needs to be chosen carefully as we require a gain of around 20dB at the test frequencies and a maximum output level of +20dBm or more. One of the best value choices is an amplifier based on the Mini Circuits PGA-103+ MMIC amplifier chip. This has a frequency range of 50kHz to 4GHz, a gain of more than 20dB to 400MHz and a maximum output power of +20dBm when used with a 5V supply. The amplifier is often sold as a ready built low-noise preamp module, such as the popular LNA4ALL.

In addition to the amplifier, you will need a 20dB 50Ω attenuator. You can use a ready-made unit or build your own with a few resistors. I've shown a schematic in **Fig.2-22**.

Most of the MMIC amplifiers, including the PGA-103+ have a gain that varies with frequency, so we need to compensate for that by running a partial calibration of the NanoVNA. To do this, connect the NanoVNA, attenuator and amplifier as shown in **Fig.2-23**.

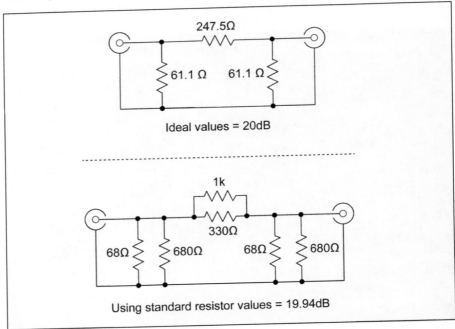

Fig. 2-22: Schematic of a 20dB attenuator.

# Measuring RF Switch / Relay Performance

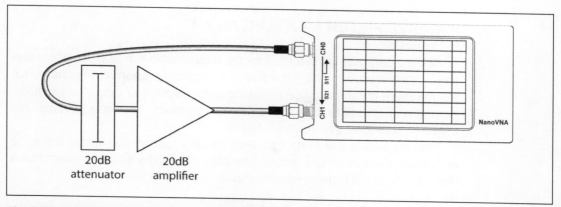

Fig. 2-23: NanoVNA with amplifier and attenuator connected.

Use the following sequence to reset the THROUGH calibration:

Starting from the top menu:
CALIBRATE – CALIBRATE – THRU wait for the tick to appear – DONE.

If all has gone to plan, you should see a flat line at 0dB on the NanoVNA, as shown in **Fig.2-24**. If you now remove the 20dB attenuator you will increase the test signal level by 20dB which will improve the dynamic range of the leakage test by 20dB. You may still find that the noise is the limiting factor but you will have gained an extra 20dB of measurement range. *NB:* A word of warning: **do not connect the amplifier output to either of the NanoVNA ports as the high level may cause permanent damage.**

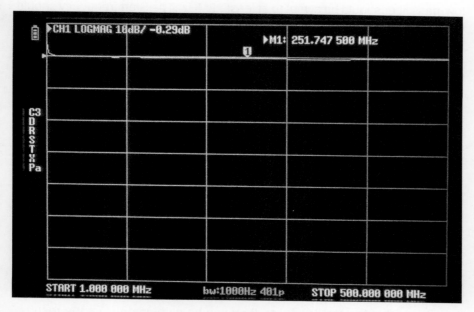

Fig. 2-24: Forward gain plot with the amplifier and attenuator in circuit.

## PASSIVE FILTER MEASUREMENT

### Introduction

A VNA is a particularly useful tool for the measurement of all types of passive filter. In addition to plotting the filter's frequency response, the NanoVNA will measure the impedance matching at the input and output of the filter. Having an immediate view of the filter's frequency response is especially useful for fine-tuning home-brew filters.

I've separated the tests between passive and active filters because additional precautions are required for active filters (by 'active', I mean filters that include some form of amplification).

### Connections

You will be using both NanoVNA ports for the filter measurements, so you will need a pair of test leads. For this type of measurement, I find 12in RF test cables the most versatile. I have tried 6in leads in the past but I found this often puts too sharp a bend at the VNA end of the lead. These comprise a standard SMA plug at one end and either SMA, BNC or N-Type plug at the other.

Always buy good quality leads and I've found that Amphenol leads are particularly good value. You will also need a through connector so the ends of the test cables can be joined together for the Through calibration.

### Configuration

The first step is to set the sweep range. This should be set to slightly higher than the *third harmonic* of the filter cut-off frequency. For example, a 7MHz low-pass filter should be swept to at least 22MHz. The Start frequency should be set to the lowest frequency you expect the filter to handle. For my 7MHz low-pass filter, I set the start at 50kHz and stop at 22MHz.

Here are the menu steps to configure the NanoVNA to measure the forward loss and the return loss at the input:

From the top menu:

STIMULUS – START – Enter desired start frequency – STOP – Enter desired stop frequency;
DISPLAY – TRACE – TRACE 0 (click till ticked);
BACK – FORMAT – LOGMAG;
BACK – CHANNEL – CH1 THROUGH;
BACK – BACK – DISPLAY – TRACE – TRACE 1 (click till ticked);
BACK – FORMAT – LOGMAG – CHANNEL – CH0 REFLECT.

### Calibration

As we're using both NanoVNA ports, we need to include the Isolation and Through calibration steps where provided. The calibration should be done at the filter end of the test leads, so they can be eliminated from the results.

# Filter Measurement

When doing the Isolation calibration step, remember to terminate the test cable from both ports with a 50Ω load.

From the top menu:

CALIBRATE – RESET – CALIBRATE;
Connect the open circuit load and press OPEN – wait for the tick;
Connect the short circuit load and press SHORT – wait for the tick;
Connect the 50Ω load and press LOAD – wait for the tick;
With the load still connected press ISOLATION – wait for the tick (*NB:* not all firmware has this step);
Connect CH0 to CH1 – press THROUGH – wait for the tick – DONE;
Choose a memory location (SAVE 0 – SAVE 5) for the calibration;
Click on the main screen to close the menu.

## Measurement

To make the display easy to read, I suggest you start with just two traces; one for the through loss that will show the filter's frequency response and the second trace to show the return loss at the filter input (though you can use VSWR if you prefer). This will show the quality of the matching at the filter input.

I've shown an example trace from a 7MHz low-pass filter in **Fig.2-25**. To see a more detailed view of the input matching, enable a third trace and set the Channel to CH0 and the format to Smith. This will display a Smith chart where you should see the trace close the central 50Ω point for the pass

Fig. 2-25: 7MHz low-pass filter response.

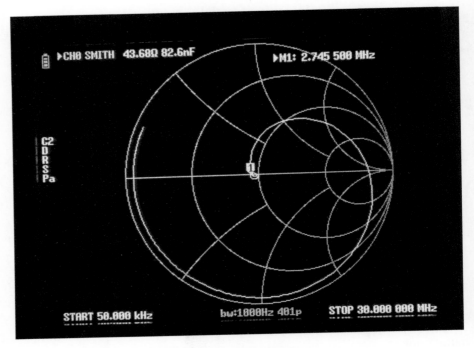

Fig. 2-26: 7MHz low-pass filter plotted on Smith chart.

band of the filter and then moving to the outer circle outside the passband as shown in **Fig.2-26**.

Once you're happy with the forward measurements, you can move on to the reverse measurement to see the frequency response in the opposite direction and check the output matching quality. To do this, leave the NanoVNA settings as they are but simply reverse the test leads, so that CH0 is connected to the filter output and CH1 to the filter input. Many simple filter designs with show a very similar result when operated in the reverse direction. You can also try using the Smith chart as described in the forward measurement.

## MEASURING ACTIVE FILTERS / AMPLIFIERS

In this context, an *active* circuit is any circuit that includes some form of amplification, so it has the potential to be overloaded by the test signal from the NanoVNA – or to overload the NanoVNA input stages.

We therefore need to think carefully about the signal levels used for testing, or the results will be compromised. This situation is further complicated because the low-cost, basic, NanoVNA does not include an output level control. Also, the higher frequency ranges of the NanoVNA use harmonics of the square-wave test signal. That means that the much lower frequency but stronger fundamental signal will still be present in the test signal.

# Filter Measurement

Please see the NanoVNA Quirks section on page 9 for a more complete explanation of the use of harmonics. The standard output level from the NanoVNA is a square wave of approximately 600mV peak-peak, which equates to 300mV RMS or +2.5dBm in 50Ω. The only practical option for controlling the NanoVNA output is to fit an attenuator between CH0 and the device under test. The attenuation value will depend on the permissible input and output levels of the device under test and the maximum input to the NanoVNA CH1.

Fig. 2-27: Mini-Circuits ZFL-500HLN amplifier.

By way of an example, I have a couple of Mini-Circuits ZFL-500HLN low-noise amplifiers, **Fig.2-27**, that I use as general-purpose RF amplifiers in the shack. These provide a very flat 20dB of gain from 10MHz to 500MHz and have a maximum output of +16dBm with 1dB compression. When testing with the NanoVNA, I need to control the input level so that the output is kept safely below the +16dBm limit, but I also need to ensure that it doesn't exceed the maximum input level of the NanoVNA, which is +10dBm. The easiest solution for most situations is to add an attenuator with an attenuation value that matches the gain of the amplifier. In this example, that would be a 20dB attenuator to offset the expected 20dB gain of the amplifier.

The attenuator will drop the input level to the amplifier from +2.5dBm without the attenuator to −17.5dBm. The amplifier output will similarly drop to +2.5dBm, comfortably below the +10dBm maximum of the NanoVNA input, **Fig.2-28**.

The snag with this configuration is that you can't measure the amplifier input matching because it is masked by the large (20dB) attenuator.

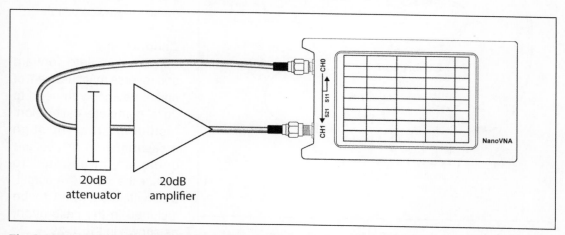

Fig. 2-28: Amplifier and gain matching attenuator.

**NanoVNA V2:** This new model avoids harmonics and uses a lower output level of approximately −10dBm. There is also a four-step level adjustment that can further reduce the output level to −20dBm. As a result, you don't have to worry about amplifier overload from the fundamental. However, you *will* still need to take care with the input level to your filter or amplifier to ensure the signal presented to CH1 is safely below the +10dBm upper limit.

**Attenuator Tip:** In most cases, you can avoid the maths by simply using an attenuator with the same value as the gain of the active device, i.e. use a 20dB attenuator with a 20dB amplifier.

## Configuration

We'll use TRACE 0 for the gain measurement in dB and TRACE 1 to show the input matching. Here are the menu steps to configure the NanoVNA:

STIMULUS – START – Enter start frequency;
STOP – Enter stop frequency;
BACK – DISPLAY – TRACE – select TRACE 0;
BACK – FORMAT – LOGMAG;
BACK – CHANNEL – CH1 THROUGH;
BACK – TRACE 1
BACK – FORMAT – LOGMAG;
BACK – CHANNEL – CH0 REFLECT
BACK – SCALE – REFERENCE POSITION – 2 – BACK.

*NB:* The SCALE change is necessary because, once we complete the calibration the amplifier gain will be positive at around +20dB which would be above the default scale. Changing the SCALE reference position makes the trace visible.

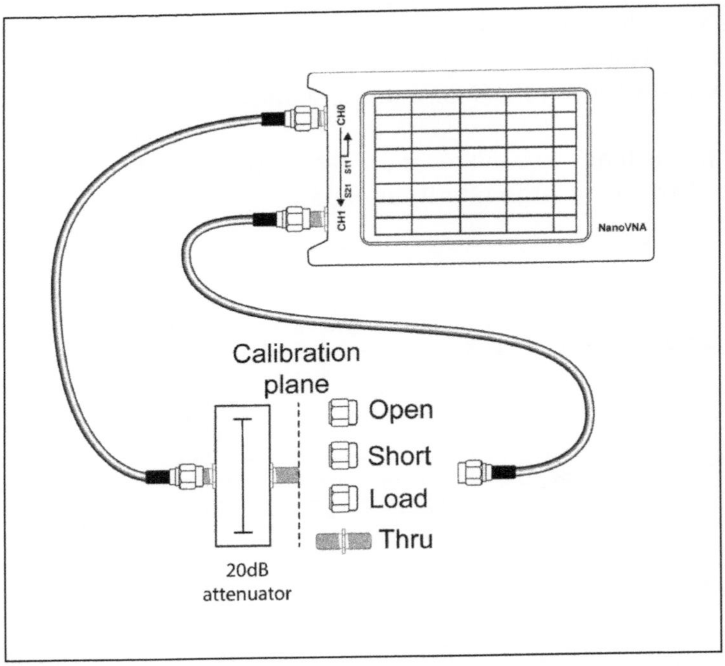

## Calibration

For the complete forward and reverse measurement of active devices you need to do a complete calibration, including the through measurements. If you are using an attenuator to reduce the NanoVNA output, that will also need to be included in the calibration, as shown in **Fig.2-29**.

**Fig. 2-29: Calibrating the amplifier test configuration.**

# Measuring Attenuators

Calibration steps:

From the top menu:
CALIBRATE – RESET – CALIBRATE;
Connect the open circuit load and press OPEN – wait for the tick;
Connect the short circuit load and press SHORT – wait for the tick;
Connect the 50Ω load and press LOAD – wait for the tick;
With the load still connected press ISOLATION – wait for the tick (*NB:* not all firmware has this step);
Connect CH0 to CH1 – press THROUGH – wait for the tick – DONE;
Choose a memory location (SAVE 0 – SAVE 5) for the calibration;
Click on the main screen to close the menu.

## Measurement

To make the measurement connect the NanoVNA, amplifier and attenuator as shown in Fig.2-28. As we included the attenuator in the calibration the NanoVNA will report the amplifier's corrected gain, i.e. +20dB in the case of my Mini Circuits amplifier.

# MEASURING ATTENUATORS
## Introduction

Attenuators are simple passive devices that are often overlooked when it comes to equipping an RF workshop. However, good quality attenuators have a myriad of uses in RF test and measurement, **Fig.2-30**.

The most useful attenuators exhibit a constant, accurately defined, attenuation over a wide frequency range. You will find plenty of second-hand examples available from popular auction sites. Whist building accurate attenuators for microwave can be challenging, HF and VHF attenuators are well within the capabilities of most amateurs. The NanoVNA is a vital tool in attenuator construction as it facilitates very accurate attenuation and impedance matching measurement.

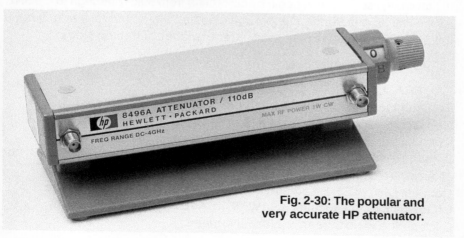

Fig. 2-30: The popular and very accurate HP attenuator.

# NanoVNAs Explained

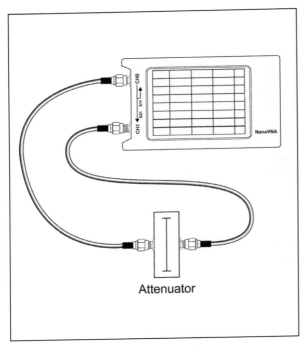

Fig. 2-31: Attenuator testing connection.

## Connections
You will need a pair of test leads with the appropriate connectors for attaching NanoVNA CH0 to the attenuator input and CH1 to the output, **Fig.2-31**.

## Settings
I suggest you configure the NanoVNA for through loss from CH0 to CH1, Return loss on CH0 plus a third trace with a Smith chart on CH0.

From the top menu:

STIMULUS – START – Enter start frequency;
STOP – Enter stop frequency;
DISPLAY – TRACE – select TRACE 0
BACK – FORMAT – LOGMAG;
BACK – CHANNEL – CH1;
DISPLAY – TRACE – select Trace 1;
BACK – FORMAT – LOGMAG;
BACK – CHANNEL – CH0;
DISPLAY – TRACE – select Trace 2;
BACK – FORMAT – SMITH;
BACK – CHANNEL – CH0.

## Calibration
Before starting the calibration, you need to decide the frequency range that you want to check. If using a commercial attenuator you can use the manufacturer's specifications but for a home-brew design you can use the NanoVNA measurements to establish the workable range of your attenuator. In that case, I suggest you begin by calibrating from 50kHz through to 1.5GHz. As we will be measuring the through loss and the impedance matching, a full SOLT calibration is necessary. Here are the calibration steps:

From the top menu:

CALIBRATE – RESET – CALIBRATE;
Connect the open circuit load and press OPEN – wait for the tick;
Connect the short circuit load and press SHORT – wait for the tick;
Connect the 50Ω load and press LOAD – wait for the tick;
With the load still connected press ISOLATION – wait for the tick (*NB:* not all firmware has this step);
Connect CH0 to CH1 – press THROUGH – wait for the tick – DONE;
Choose a memory location (SAVE 0 – SAVE 5) for the calibration;
Click on the main screen to close the menu.

## Measurement

Attenuator measurement is straightforward as you simply connect CH0 to the input and CH1 to the attenuator output. Trace 0 will show the through loss of the attenuator (S21) whilst Trace 1 will indicate the input matching quality (S11). Trace 2 can also be studied to view the input matching S(11) on a Smith chart. If your attenuator is designed to operate in a 50Ω system, you can reverse the input and output connections to check the output matching (S22).

# DIRECTIONAL COUPLERS

Directional couplers are used extensively in amateur radio for SWR measurement, PA protection control and anywhere a directional sample is useful. These devices also make an easy home-brew project, but the result can be far more useful if the coupler's performance can be accurately measured. The NanoVNA is the ideal instrument for this measurement, as I will show you here.

The purpose of a directional coupler is to sample a signal in a given direction. For example, to measure the standing wave ratio on a transmission line we need two directional couplers. A forward coupler to sample the output from the transmitter (forward sample), so we know the level of the signal being sent into the transmission line, and another coupler that samples the reflected signal to indicate the level of the reflected signal that's caused by a mismatch in the transmission line. I've shown the signal flows in **Fig.2-32**. Some simple maths on the forward and reflected samples can provide the VSWR (Voltage Standing Wave Ratio).

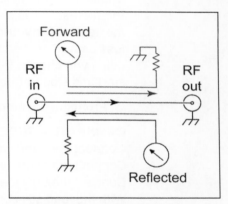

**Fig. 2-32: Signal flows through a directional coupler.**

## Connections

To measure the forward and reflected performance of the coupler you will need two test leads with the appropriate connectors to match the coupler. You will also need one or two 50Ω termination loads that match the coupler connections. This is necessary because any unconnected port must be terminated to achieve accurate results.

## Settings

As we will be measuring the matching quality and the loss between various ports, I suggest you configure Trace 0 for forward loss (S21), Trace 1 for input matching (S11) in return loss and Trace 2 also for input matching (S11) but using a Smith chart. Here is the menu sequence for that configuration:

From the top menu:

STIMULUS – START (set start frequency) – STOP - (set stop frequency);
BACK - DISPLAY – TRACE – TRACE 0
BACK – FORMAT – LOGMAG
BACK – CHANNEL – CH1 THROUGH

Trace 1: From top menu:
DISPLAY – TRACE – TRACE 1;
BACK – FORMAT – LOGMAG;
BACK – CHANNEL – CH0 REFLECT.

Trace 2: From top menu:
DISPLAY – TRACE – TRACE 2;
BACK – FORMAT – SMITH;
BACK – CHANNEL – CH0 REFLECT.

**Calibration**

The first step in the calibration process is to determine the frequency range for the tests. This should be selected to be approximately 10% wider than the frequencies you intend to pass through the coupler. As we'll be measuring loss as well as matching you will need to complete a full SOLT calibration using you chosen test leads. Use the following menu sequence to set the start and stop frequencies:

From the top menu:

CALIBRATE – RESET – CALIBRATE;
Connect the open circuit load and press OPEN – wait for the tick;
Connect the short circuit load and press SHORT – wait for the tick;
Connect the 50Ω load and press LOAD – wait for the tick;
With the load still connected press ISOLATION – wait for the tick (*NB:* not all firmware has this step);
Connect CH0 to CH1 – press THROUGH – wait for the tick – DONE;
Choose a memory location (SAVE 0 – SAVE 5) for the calibration;
Click on the main screen to close the menu.

**Measurement**

The first important point to note is that you will only see the coupler's designed performance when all ports are properly terminated, normally 50Ω. During the testing, it's essential to terminate any unconnected ports.

The first measurement is the through, or insertion loss between the main input and output (S21) of the coupler, **Fig.2-33**. Start by fitting terminating loads to the coupler port(s). Connect CH0 to the coupler input and CH1 to the output. A good quality coupler will exhibit a very low loss of a fraction of a dB and a return loss of 30dB or better.

Next, we'll measure the coupling in the forward direction. Leave CH0 connected to the input but move CH1 to the forward coupler output and

# Directional Couplers

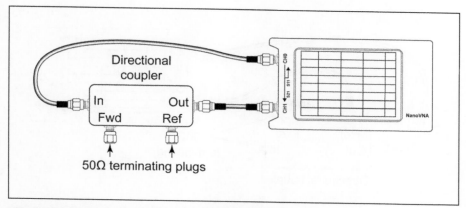

**Fig. 2-33: Directional coupler through measurement.**

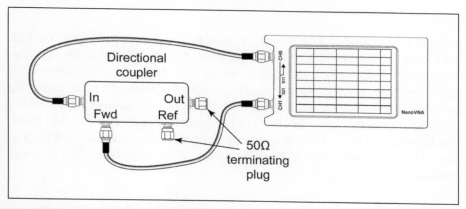

**Fig. 2-34: Forward coupling measurement.**

apply a terminating load to the main output and the reverse coupler port, **Fig.2-34**. In this case you will measure the forward coupler loss, which may well be around 20dB, so the NanoVNA will indicate −20dB. For a well-designed coupler, the loss will be constant over the operating frequency range.

Next, we swap coupler ports, connecting CH1 to the reverse coupler port and connecting a terminating load to the forward coupler port. You are now measuring the directivity of the coupler. In a perfect coupler, you would see no signal at the reverse port. However, there will be a small reflected wave on the line in a practical design, and the directivity will be compromised by stray capacitance and inductance. You will see a reading of −40dB or better on the reverse port in a well designed coupler. This reading will almost certainly be frequency dependent.

At this point, we know the forward loss, forward coupling and reverse coupling directivity. Next, we need to measure the reverse coupling and forward directivity. Connect CH0 to the main coupler output and connect a terminating load to the main input. Connect CH1 to the reverse coupler

# NanoVNAs Explained

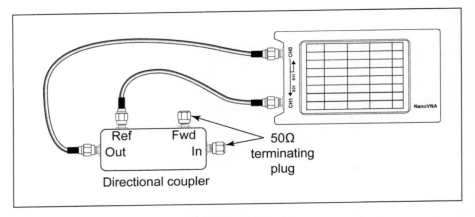

**Fig. 2-35: Reverse coupling measurement.**

output and connect a terminating load to the forward coupler output, **Fig.2-35**. In this case we're sending the test signal in the reverse direction through the coupler and measuring the coupling to the reverse port. A well-designed coupler will deliver an output of around −20dB with a flat frequency response. The final measurement is to check the directivity of the forward coupler. To do this, connect CH1 to the forward coupler output and move the terminating load to the reverse coupler port. The response here should be −40dB or better for a 20dB coupler. That completes the coupler measurements.

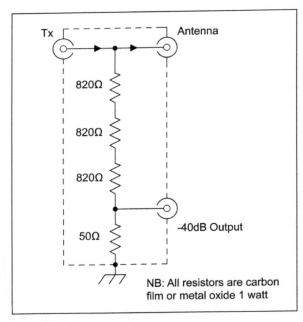

**Fig. 2-36: RF tap schematic.**

## RF TAP
### Introduction
An RF tap or 'sniffer' is a simple resistive tap of an RF signal that's used for signal monitoring. A common value in amateur radio is a 40dB tap. The purpose of the tap is to extract a low-level sample of the transmitted signal so that it can be monitored with a spectrum analyser or 'scope. A 40dB tap attenuates a 100-watt signal down to around 700mV RMS which is ideal for connecting to test equipment. I've shown a typical RF tap circuit in **Fig.2-36**. As you can see it's a three-port device, with an input, output and the tap. If we know the precise attenuation of the tap we can use the output to measure the transmit power into a 50Ω load.

90

# RF Tap

## Connections

You will need test leads with connectors to fit the three ports of the tap along with a terminating load. As with the directional coupler, you will only be able to measure the design performance if all ports are correctly terminated.

## Settings

The first step is to decide the frequency range and for a home-brew project that is simply the range of frequency you intend to operate using the coupler. If you're testing a commercial product then you could use the manufacturer's specified operational range.

From the top menu:

STIMULUS – START (set start frequency) – STOP – (set stop frequency);
BACK – DISPLAY – TRACE – TRACE 0 (click to show the tick);
BACK – FORMAT – LOGMAG (click to show tick);
BACK – CHANNEL – CH1 THROUGH (click to show tick);
BACK – BACK – DISPLAY – TRACE – TRACE 1 (click to show tick);
BACK – FORMAT – LOGMAG;
BACK – CHANNEL – CH0 REFLECT (click to show tick).

Trace 2: From the top menu:

DISPLAY – TRACE – TRACE 2;
BACK – FORMAT – SMITH;
BACK – CHANNEL – CH0 REFLECT (click to show tick).

## Calibration

As we require through measurements we need the full SOLT calibration to the ends of the test leads.

From the top menu:

CALIBRATE – RESET – CALIBRATE;
Connect the open circuit load and press OPEN – wait for the tick;
Connect the short circuit load and press SHORT – wait for the tick;
Connect the 50Ω load and press LOAD – wait for the tick;
With the load still connected press ISOLATION – wait for the tick NB: Not all firmware has this step;
Connect CH0 to CH1 – press THROUGH – wait for the tick – DONE;
Choose a memory location (SAVE 0 – SAVE 5) for the calibration;
Click on the main screen to close the menu.

## Measurement

We only require two measurements to check the RF tap. The first is the through loss from input to output. This should be as low as possible and typically less than 0.5dB. The second measurement is the loss from the

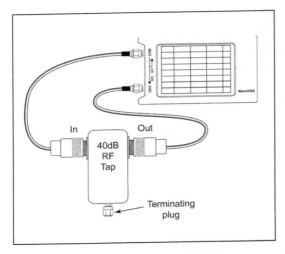

Fig. 2-37: RF tap forward measurement.   Fig. 2-38: RF tap, tap loss.

input to the tap output, which should be 40dB for a 40dB tap design.

For the through measurement, connect CH0 to the main input, CH1 to the main output and connect a terminating load to the tap port, **Fig.2-37**. The resulting display should show a very low loss on CH0 of better than 0.5dB and a flat frequency response over the operating range. The return loss should be higher than 20dB. To measure the tap loss, move the CH1 connection to the tap output and fit a terminating load to the main output, **Fig.2-38**. Trace 0 should show the design loss, i.e. –40dB for a 40dB tap and the return loss should be around 20dB or higher.

## COMMON MODE CHOKES
### Introduction
Common mode chokes are used extensively in all areas of electronics. Their purpose is to block the flow of unwanted longitudinal signals through all manner of cables. The most common applications are to be seen in the ferrite chokes on computer leads and the chokes used to prevent radiation and interference problems on antenna feeders. The full and precise measurement of a common mode choke's performance is a complex subject and one that has sparked many papers with alternative measurement methods. A much simpler approach can be taken if we're prepared to forego precision. In many amateur radio use cases we're creating our own chokes often using materials from the junk box or those bought online or at rallies. In these cases, we will be experimenting and need to know if a home-brew choke is providing a reasonable choking effect in the desired frequency range.

I can offer you two measurement solutions. The first is a simple but crude measurement that is useful for checking the approximate choking

# Common Mode Chokes

effect at the desired frequencies. The second method is adapted from an article by Steve Hunt, G3TXQ, and published in *RadCom Plus* in May 2015.

## Method 1

For this method, we connect the Common Mode choke between CH0 of the NanoVNA and ground, as shown in **Fig.2-39**. We then measure the resistive impedance of the choke over the frequency range of interest. As the impedance of an effective choke is likely to be several thousand ohms at its operating frequencies, the NanoVNA measurement accuracy will be compromised. This is because the NanoVNA is essentially a 50Ω measuring device. The other problem arises from the stray capacitance of the test setup as this can significantly move the resonant frequency of high-Q common mode chokes. However, the results are still useful for the experimenter as they will give a good indication of the frequency range over which the choke could be useful.

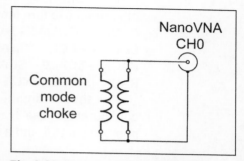

Fig. 2-39: Common mode choke connection.

## Connections

You will need to make a simple test fixture with an SMA socket and a couple of crocodile clips. Common together each side of the choke and connect it between the crocodile clips as shown in **Fig.2-40**.

## Settings

Begin by setting the Stimulus START and STOP to the frequency range you want to check. Next we configure a single trace to show the resistive element of the impedance as this is what matters for a common mode choke. As we will be measuring higher impedances than normal, we also need to change the NanoVNA scaling to get a sensible display. Here are the menu steps for this configuration:

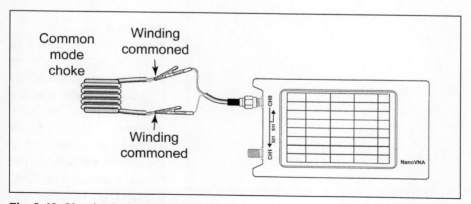

Fig. 2-40: Simple choke test setup.

From the top menu:

STIMULUS – START (set start frequency) – tap screen – STOP (set stop frequency) – tap screen;
BACK – DISPLAY – TRACE – TRACE 0 (click to show the tick and click other traces to turn them off);
BACK – FORMAT – MORE – RESISTANCE – tap screen;
BACK – BACK –- CHANNEL – CH0 REFLECT (click to show tick);
BACK – SCALE – SCALE/DIV – 2000;
BACK – REFERENCE POSITION 1.

For those familiar with the NanoVNA menu, here's a summary configuration: use Trace 0 measuring resistance with the scale reference position set to 1 and a scaling of 2000 ohms/division.

## Calibration

The simple method uses CH0 so you only need do the open / short / load calibration and save the results.

From the top menu:

CALIBRATE – RESET – CALIBRATE;
Connect the open circuit load and press OPEN (wait for the tick to appear);
Connect the short circuit load and press SHORT (wait for the tick);
Connect the 50Ω load and press LOAD (wait for the tick) – DONE;
Choose a memory location to save the calibration;
Click on the main screen to close the menu.

## Measurement

With the choke connected as shown in Fig.2-40, a working choke will likely produce a curve similar to **Fig 2-41**. To be effective in a 50Ω system, the common mode choke should have an impedance of 2000Ω or more at the operating frequencies.

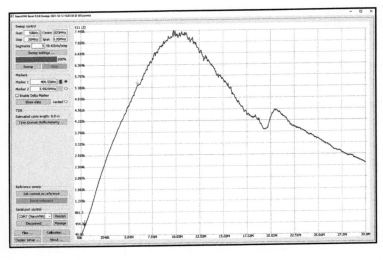

Fig. 2-41: Response of a shunt connected choke.

## Method 2

The second method is used to obtain a more accurate measurement of a common mode choke's performance. In this method, the choke windings are commoned and connected between the centre conductor of CH0 and CH1, i.e. in series with the test signal as shown in

# Common Mode Chokes

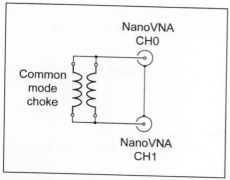

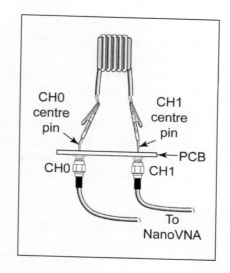

Above: Fig. 2-42: Measuring a common mode choke using a series connection.

Right: Fig. 2-43: Choke test fixture.

**Fig.2-42**. Using this configuration, we can measure the loss between CH0 and CH1 and use the S21 values to calculate the impedance of the choke. The original article employed a spreadsheet to calculate the result but we can make use of the NanoVNA Saver software to plot the results directly.

## Connections
For this method, you will need to build a simple test jig, that comprises a strip of copper, brass or PCB with two SMA panel sockets. On the socket to be used with CH0 connect a 50Ω or 2 x 100Ω film resistors to ground. This is necessary to tame the very high impedances being presented to CH0. To both centre pins, attach a short wire to a crocodile clip. When taking measurements, the choke being tested is connected between the two crocodile clips as shown in **Fig.2-43**. The purpose of the fixture is to provide a simple connection method that adds a minimum of stray capacitance.

## Calibration
With NanoVNA Saver running and connected to your NanoVNA, begin by setting the desired frequency range. Next, carry out a full SOLT calibration to the end of the test leads that you will be using. The NanoVNA Saver Calibration Assistant will guide you through this process but there is more information in the NanoVNA Saver section of this book.

## NanoVNA Saver Configuration
I suggest you use two graph displays to show the performance results. The top chart will show the forward loss in dB (S21) and will indicate the signal loss through the choke. The second plot will show the calculated series resistive and reactive (R+jX) values of the S21 sweep. In this plot, it is the resistive value that's of interest. Here are step-by-step instructions for

configuring NanoVNA Saver:
1. Start by opening Display setup panel (bottom left of NanoVNA Saver main screen);
2. Make sure Show lines is ticked;
3. Towards the top left set the Sweep colour to blue by clicking on the colour patch;
4. Set the Second sweep colour to red;
5. Set the point size to 2 px and Line thickness to 3px;
6. In the Displayed charts box set the top left graph to S21 Gain and the graph below it to S21 R+jX series.

### Measurement

With the configuration complete, you can begin measurement. Connect the choke to the test fixture as shown in Fig.2-43. Press the sweep button to display the results. If you've used the colours and layout suggested in my configuration steps, the top chart will show the loss being introduced by the choke in dB, **Fig.2-44**. However, much of this will be due to reactive losses but it's the resistive loss that's the most important for common mode chokes. The resistive component can be seen in the red trace on the lower display. Using the scale on the left you can read the resistive values which should ideally be greater than 2000Ω over the target frequency range. The blue trace shows the series reactive elements and will probably have positive and negative values as shown by the scale on the right of the graph. If you need a more detailed sweep you can use the Segments box in the sweep control section to multiply the number of measurement points.

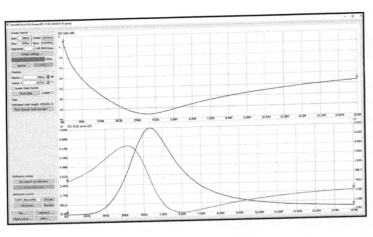

Fig. 2-44: S21 and R+jX plots of a common mode choke.

## BALUNS

### Introduction

These versatile devices are used extensively for connecting an unbalanced (coaxial) feeder to a balanced (e.g. dipole) antenna. Their primary role is to prevent common-mode currents from entering the coax screen and causing RFI. In addition to providing a balanced to unbalanced conversion, many baluns also provide impedance matching. The common-mode properties of

# Baluns and Ununs

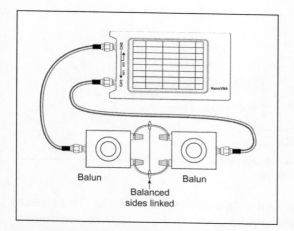

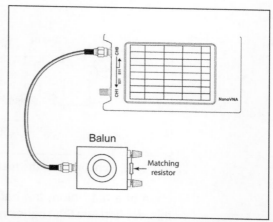

Fig. 2-45: Baluns connected back-to-back.

Fig. 2-46: Using a non-inductive resistor to check balun matching.

these important everyday components are not the easiest to measure with a VNA, because the NanoVNA, like most VNAs, is an unbalanced measuring instrument with a common ground connection for the signal return path. In addition to the important common-mode performance, we also need to be able to measure impedance matching and the balun's forward loss over the desired band. By far the simplest measurement technique is to build two identical baluns and connect them back-to-back as shown in **Fig.2-45**. In that configuration, we can check the forward loss by measuring between the two unbalanced ports and dividing the result by two. You can also check the impedance matching of a balun quite simply by terminating the balanced side with a non-inductive resistor of the appropriate value and measuring the return loss or SWR at the 50Ω unbalanced port, **Fig.2-46**. Where a balun is to be used for transmission, the power handling capacity of the core is critical but this can't be measured with a VNA.

## Common Mode Rejection

For measuring the common mode rejection, I suggest using the method devised by Ron Skelton, W6WO, that was published in the November / December 2010 issue of *QEX*. This method treats the balun as a three-port device as shown in **Fig.2-47**. The 50Ω unbalanced input forms one port whilst the other two ports are derived from the balanced side of the balun. With this technique, we need to make sure the balanced side is properly matched and this is achieved by adding a precision resistor in series with each of the two balanced legs. For example, in the

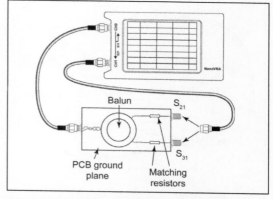

Fig. 2-47: Test fixture with series matching resistors.

# NanoVNAs Explained

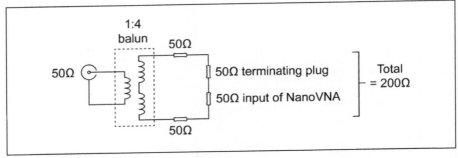

Fig. 2-48: Schematic of the test fixture.

case of a 4:1 balun, the balanced side expects to see 200Ω. To achieve this, we add two 50Ω series resistors, as shown in Fig.2-47. When combined with the NanoVNA impedance of 50Ω and a 50Ω terminating plug in the unused leg we get 200Ω, **Fig.2-48**. To make the measurements, you will need to build the simple test fixture as shown in Fig.2-47.

## Settings

As we need to save and manipulate S parameter data, I strongly recommend connecting the NanoVNA to a computer running NanoVNA Saver software. For our measurements, we need to capture the forward loss from the 50Ω port to each of the newly created secondary ports. Here are the NanoVNA Saver settings:

1. Set the Sweep start and stop to the desired band;
2. Leave Segments at 1;
3. Open the Display setup panel;
4. In the Displayed charts section, set the top left to S21 Gain and the bottom left chart to S11 Return Loss. All other charts set to None.

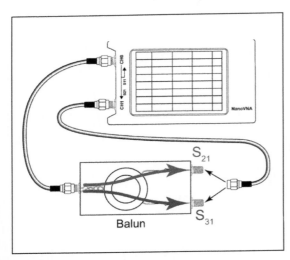

Fig. 2-49: Balun S parameters.

## Calibration

Use the Calibration Assistant in NanoVNA Saver to perform a full calibration to the end of the test leads that will be used with your test fixture.

## Measurement

As the test fixture converts the balun into a three-port device, the S parameters we require become S21 and S31 respectively, **Fig.2-49**. However, the NanoVNA remains a two-port device, so we begin by measuring the forward gain to one secondary port and save the data file as S21, then we measure to the second port and save the data file as S31. Here's a step-by-step guide:

# Baluns and Ununs

1. Connect NanoVNA CH0 to the 50Ω side of the balun;
2. Connect CH1 to one of the secondary ports on the test fixture;
3. Connect a 50Ω terminating plug to the other secondary port;
4. Start a single sweep of the band;
5. If all is well, the S21 Gain (dB) plot should show a loss of between 5 and 10dB;
6. To save the result, go to the Files menu (bottom-left) and select Save 2Port file (S2P) and choose a suitable location and include S21 in the file name;
7. Reverse the secondary connections by moving CH1 to the other secondary port and moving the terminating plug to the first;
8. Start a single sweep of the band;
9. As with the previous sweep you will see a loss of between 5 and 10dB;
10. Repeat the save process as per step 6 but include S31 in the file name.

Ron Skelton's formula for Common Mode Rejection Ratio in dB is:

$$CMRR = 20 \times log(S_{21}+S_{31})/(S_{21}-S_{31})$$

Whilst this is a simple formula, the S21 and S31 values are complex numbers, so the maths is also rather complex. To simplify the measurement, I have produced a spreadsheet that automates the calculations and produces both the CMRR and the phase balance in degrees at each test frequency. The spreadsheet is available for download from the Files section of the NanoVNA Explained user group at Groups.io.

## Impedance Matching

The simple way to check the impedance matching is to use a single trace to measure the return loss or VSWR with a non-inductive resistor of the appropriate value connected to the balanced side of the balun, Fig.2-46.

## Settings

Here are the settings to use the NanoVNA to check the balun matching. You can substitute LOGMAG with SWR or SMITH if you prefer.
STIMULUS – START (set start frequency) – tap main screen – STOP (set stop frequency);
BACK – DISPLAY – TRACE – TRACE 0 (click to show the tick);
BACK – FORMAT – LOGMAG (click to show tick)
BACK – CHANNEL – CH0 REFLECT;
BACK – tap main screen to finish.

## Calibration

You need to run a Short-Open-Load calibration to the end of the test lead that will connect to the balun as follows:
CALIBRATE – RESET – CALIBRATE;
Connect the open circuit load and press OPEN – wait for the tick;
Connect the short circuit load and press SHORT – wait for the tick;

Connect the 50Ω load and press LOAD – wait for the tick ;
Click DONE;
Choose a memory location (SAVE 0 – SAVE 5) for the calibration;
Click on the main screen to close the menu.

**Measurement**
Connect a non-inductive resistor of the appropriate value to the high impedance side of the balun and connect NanoVNA CH0 to the 50Ω port. You will now see a trace showing the return loss (or SWR / Smith) that will show the quality of matching over the selected frequency range.

# UNUNS

## Introduction
Whilst a balun is used to connect an unbalanced (usually coax) feeder to a balanced antenna, an *unun* has an unbalanced device on both sides, hence the name UNbalanced to UNbalanced. By far the most common use of an unun is to connect a coax feeder to a high impedance, unbalanced antenna such as a random wire or a vertical.

Although the unun description has appeared fairly recently it has a close cousin that has been in use since the early days of radio: the auto-transformer. The distinguishing feature of both devices is the use of a single tapped coil instead of two or more isolated windings, **Fig.2-50**.

Whilst most ununs are constructed using three windings on the same core, they are joined together to create a single winding with no DC isolation. As you can see, the unun has a common ground line along with a low Z primary and a higher Z secondary. The most popular form of the unun provides a 9:1 impedance transformation.

Being an unbalanced device we can use the NanoVNA to measure the performance, but we do have an impedance matching problem. As with the balun, the simplest way to measure performance is to build two ununs and join them back-to-back, **Fig.2-51**. With that config-

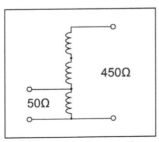

Fig. 2-50: Schematic of a typical unun.

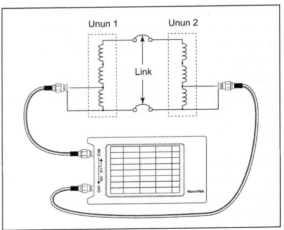

Fig. 2-51: Testing two ununs back-to-back.

# Baluns and Ununs

uration the NanoVNA will be connecting to 50Ω ports. You can also check the impedance matching by connecting the appropriate value, non-inductive, resistor to the secondary and measuring the SWR or return loss from the 50Ω port, **Fig.2-52**.

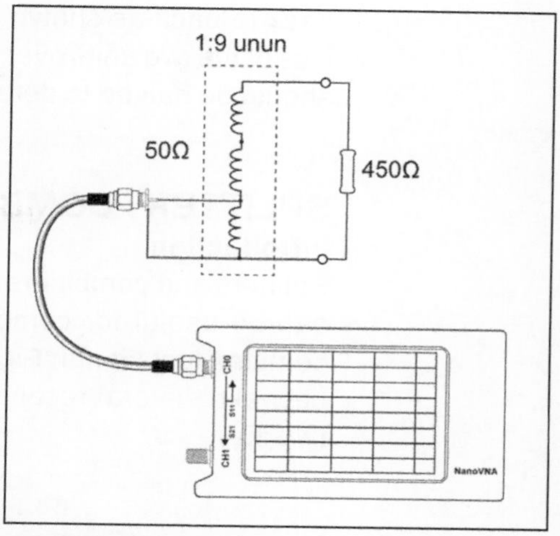

## Settings

I suggest setting the NanoVNA to measure S11 return loss and S21 forward gain using the following steps:

STIMULUS – START – enter desired start frequency. Tap main screen – STOP – enter desired stop frequency;
BACK – DISPLAY – TRACE - TRACE 0 – click to show tick;
BACK – FORMAT – LOGMAG;
BACK – CHANNEL – CH0 REFLECT
BACK – BACK – DISPLAY – TRACE – TRACE 1 – click to show tick;
BACK – FORMAT – LOGMAG;
BACK – CHANNEL – CH1 THROUGH;
BACK – BACK.

**Fig. 2-52: Resistor used to check impedance matching.**

*NB:* for TRACE 0, you can swap LOGMAG for Smith or SWR if you prefer those formats.

## Calibration

To check the impedance matching and the loss through back-to-back ununs you will need to complete a full calibration over the band of frequencies.

CALIBRATE – RESET – CALIBRATE;
Connect the open circuit load and press OPEN – wait for the tick;
Connect the short circuit load and press SHORT – wait for the tick;
Connect the 50Ω load and press LOAD – wait for the tick;
With the load still connected press ISOLATION – wait for the tick (*NB:* not all firmware has this step);
Connect CH0 to CH1 – press THROUGH – wait for the tick – DONE;
Choose a memory location (SAVE 0 – SAVE 5) for the calibration;
Click on the main screen to close the menu.

## Measurement

To measure the impedance matching, connect CH0 to the 50Ω port of the unun and a non-inductive resistor of the appropriate value (i.e. 450Ω for a 9:1 unun) to the high impedance secondary, Fig.2-52. Once connected, TRACE 0 will display the matching quality in the form of return loss.
To measure the forward loss of the unun, connect two identical ununs

back to back as shown in Fig.2-51. TRACE 1 will then show the forward loss of the two units over the selected frequency range. The measured result should be halved to determine the loss for a single unun.

## SPLITTER / COMBINER
### Introduction
Splitters and combiners are very useful accessories for any test bench and can be useful for combining signals from signal generators to create a complex test signal, **Fig.2-53**. They are also used for splitting an antenna between several receivers. When testing a splitter or combiner there are two key measurements. The first is the forward loss of the device whilst the second is the isolation between ports. The isolation is important because we don't want the impedance variations of one device to affect the performance of another. The measurement of splitter / combiner performance is straightforward, but there is one golden rule: all unused ports must be terminated (normally with a 50Ω load).

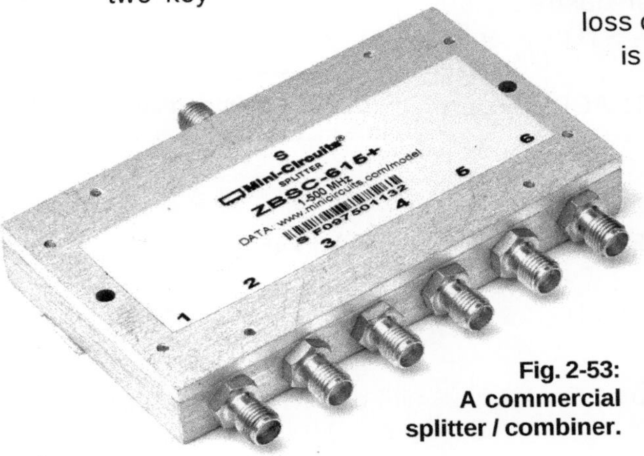

**Fig. 2-53: A commercial splitter / combiner.**

### Configuration
I suggest you use two traces, one to show the impedance match at CH0 and the other to show the loss to CH1. Here are the configuration steps:

STIMULUS – START – enter desired start frequency;
Tap main screen – STOP – enter desired stop frequency;
BACK – DISPLAY – TRACE - TRACE 0 – click to show tick;
BACK – FORMAT – LOGMAG;
BACK – CHANNEL – CH0 REFLECT;
BACK – BACK – DISPLAY – TRACE – TRACE 1 – click to show tick;
BACK – FORMAT – LOGMAG;
BACK – CHANNEL – CH1 THROUGH – BACK – BACK.

### Connections
Start with two test leads with the appropriate connectors to match the unit being tested. You will also need several 50Ω terminating plugs to terminate all the unused ports. As a simple guide, you will require one less termination than the total number of ports. For example, to test a four-port combiner you will need 3 x 50Ω loads.

# Splitter / Combiner

## Calibration

Assuming you have completed the configuration steps you need to do a full SOLT calibration as follows:

From the top menu:

CALIBRATE – RESET – CALIBRATE;
Connect the open circuit load and press OPEN – wait for the tick;
Connect the short circuit load and press SHORT – wait for the tick;
Connect the 50Ω load and press LOAD – wait for the tick;
With the load still connected press ISOLATION – wait for the tick (*NB:* not all firmware has this step);
Connect CH0 to CH1 – press THROUGH – wait for the tick – DONE;
Choose a memory location (SAVE 0 – SAVE 5) for the calibration;
Click on the main screen to close the menu.

## Measurement

Most splitter / combiners are bi-directional, so begin by connecting CH0 to the input port and CH1 to the first output port. Connect 50Ω loads to the remaining ports, **Fig.2-54**.

Now measure the forward loss, S21 as shown on TRACE 1. Repeat that measurement for each port whilst ensuring that all unused ports have a terminating load attached. The next step is to measure the isolation between ports. Begin by connecting a terminating load to the input port. Next connect CH0 to port 1 and CH1 to port 2 and measure the loss. Repeat the measurement by moving CH1 to each port in turn whilst ensuring all unused ports are terminated.

For a complete view of the splitter / combiner's performance you need to repeat this process for each port. At the end of the process you will have measured the isolation between each port and every other port. A good quality splitter / combiner should have a port isolation of 30dB or better.

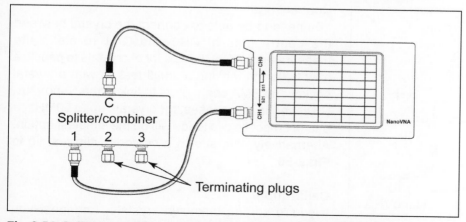

**Fig. 2-54: Splitter / combiner connection for through measurements.**

# MEASURING CRYSTALS

## Introduction

Crystal oscillators used to be the favourite way for radio amateurs to obtain frequency accuracy and stability. Crystals are also used to build high-quality RF filters. That heavy reliance on the crystal has been largely superseded by the use of TCXO (Temperature Compensated Crystal Oscillator) modules and programmable synthesisers. The widespread introduction of SDR techniques has also reduced the demand for crystal filters. However, there is still a role for the humble crystal, especially in low cost, low power rigs.

Crystals have two resonant modes known as series and parallel resonance and these occur on slightly different frequencies. When planning to use a crystal it is helpful to know the series and parallel characteristics of your selected crystal. In **Fig.2-55**, I've shown the equivalent schematic of a crystal resonator. At the series resonance point, the crystal presents a low impedance, whilst the parallel resonance creates a high impedance.

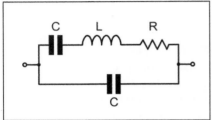

Fig. 2-55: Crystal schematic showing series and parallel resonant components.

The NanoVNA can be used to accurately measure a crystal's resonance. This can be useful when sorting miscellaneous crystals or selecting crystals to build a filter. However, there is an issue with the V2 NanoVNA and crystals. The NanoVNA V2 pulses the test signal at a rate of a few kilohertz. Whilst not normally a problem, crystals are electro-mechanical devices, i.e. the crystal vibrates, and the V2's pulsing test signal doesn't leave sufficient settling time with the signal applied. As a result, NanoVNA V2 crystal measurements will be unreliable. There has been some special firmware developed to provide a crystal measurement option and you will find that via the NanoVNA V2 user group. All the other version 1 NanoVNAs use a continuous test signal, so are ideal for use testing crystals.

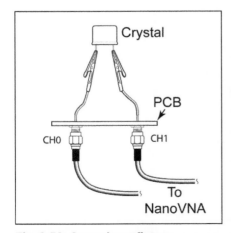

Fig. 2-56: Crystal test fixture.

## Connections

You need to be able to connect the crystal between the centre pins of CH0 and CH1 to make the measurement. If you have a lot of crystals to check, it may be worth making a small test jig with a crystal socket and SMA connectors to link to the NanoVNA. However, in most cases, for crystals up to 50MHz, a pair of crocodile clip leads will be the simplest option. Alternatively, I've shown a crocodile clip test jig in **Fig.2-56**.

## Calibration

If you're sorting unknown crystals, I suggest you calibrate the NanoVNA over the range 50kHz to

# Measuring Crystals

50MHz. We'll be using a much narrower range for the precise crystal measurement, but the NanoVNA will interpolate from this main calibration. You need to run a full SOLT calibration and the calibration load can be a simple 50Ω resistor or a combination of values to make 50Ω connected between the crocodile clips. I used three half-watt metal oxide 150Ω resistors.

From the top menu:

CALIBRATE – RESET – CALIBRATE;
Connect the open circuit load and press OPEN – wait for the tick;
Connect the short circuit load and press SHORT – wait for the tick;
Connect the 50? load and press LOAD – wait for the tick;
With the load still connected press ISOLATION – wait for the tick (*NB:* not all firmware has this step);
Connect CH0 to CH1 – press THROUGH – wait for the tick – DONE;
Choose a memory location (SAVE 0 – SAVE 5) for the calibration;
Click on the main screen to close the menu.

**Method**

Connect the test leads as shown in **Fig.2-57**, i.e. join the ground lines from CH0 and CH1 and connect the crystal between the centre pins of CH0 and CH1. Alternatively use the test jig I described earlier.

Crystal resonances are very sharp and may not be visible on the full 50kHz to 50MHz sweep, so you may need to move through the band in smaller chunks to find the resonance. In **Fig.2-58** I've shown a plot of a 13MHz crystal with start and stop frequencies of 50kHz and 50MHz respectively. As you can see, the crystal shows as a very small blip. If you really have no idea of the crystal frequency, move through the spectrum in 5MHz chunks, i.e. 50kHz – 5MHz, then 5MHz – 10MHz, etc. If you know a likely frequency, start with a 5MHz wide span centred on the expected frequency. You often find that there are several spikes near the resonance and it's usually the lowest and strongest spikes that are the true resonance, **Fig.2-59**. Once you've detected the resonance you need to narrow the sweep so that the resonance fills most of the screen. To find the revised start and

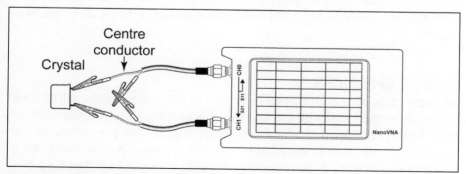

Fig. 2-57: Using a pair of crocodile clip test leads.

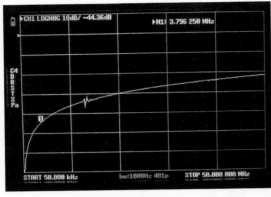 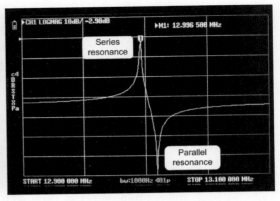

**Fig. 2-58: Wide range sweep of a crystal.**   **Fig. 2-59: Narrow sweep showing the two resonance spikes.**

stop frequencies, use the marker wheel on top of the NanoVNA to steer the marker along the trace. Set the start frequency to the point where the resonance curve just starts to rise and the stop point where the curve starts to flatten.

You can now use the marker wheel to accurately read the resonant frequencies, Fig.2-59. The first, positive peak is the *series resonant frequency*, whilst the second, negative peak is the *parallel resonant frequency*. It's important to zoom into the resonance because the relatively small number of measurement points in the basic NanoVNA could miss the true resonance peak.

### Overtone Crystals

Crystals for higher frequencies often use the third harmonic of a crystal's resonant frequency. If you know the approximate frequency of your crystal, you should begin with a 5MHz band centred on the crystal's frequency and repeatedly narrow the sweep to home-in on the resonance in the same way as with the fundamental. If the crystal frequency is unknown, start by finding the fundamental, as described earlier, and set up a new sweep at three times the fundamental frequency. If the crystal is not suitable for overtone working, you will find the third harmonic resonance to be quite weak or absent.

## CABLE CHECKER

Amateur radio stations inevitably use a lot of RF cables. Some will be used for test purposes whilst others connect the working station equipment. An intermittent cable fault can be a real problem to find but the NanoVNA can be pressed into service as a cable checker. To do this, the NanoVNA is configured to measure the forward gain (S21) between CH0 and CH1 over the range 50kHz to 250MHz. The suspect cable is then connected between

# Cable Checker

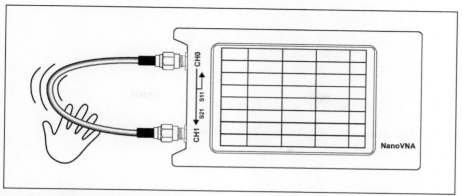

**Fig. 2-60:** Cable tester connections.

CH0 and CH1 and wiggled about whilst watching the NanoVNA trace, **Fig.2-60**. Any spikes or blips on the trace that change as the cable is moved indicate that the connection is less than perfect.

## Settings
To check the cable, I suggest setting up two traces, one to measure the through loss and the other to measure the impedance match at CH0. The frequency range is not critical, so I suggest using the full range of 50kHz to 900MHz. Here are the steps for that configuration:

From the top menu:

STIMULUS – START – enter 50k – STOP – enter 900M
BACK – DISPLAY – TRACE – TRACE 0 (click until ticked);
BACK – FORMAT – LOGMAG (click until ticked);
BACK – CHANNEL – CH 1 THROUGH
BACK – TRACE – TRACE 1 (click until ticked);
BACK – FORMAT – LOGMAG (click until ticked);
BACK – CHANNEL – CH 0 REFLECTED;
BACK – BACK.

## Calibration
There is no need to do a separate calibration for this check as we're not making critical measurements but simply looking for an intermittent loss of signal. You can use a full range calibration if you have that stored in the calibration menu Save 0 which is accessed via the RECALL menu.

# NanoVNA-H4 - Menu Map

This menu map is based on firmware version 1.0.53

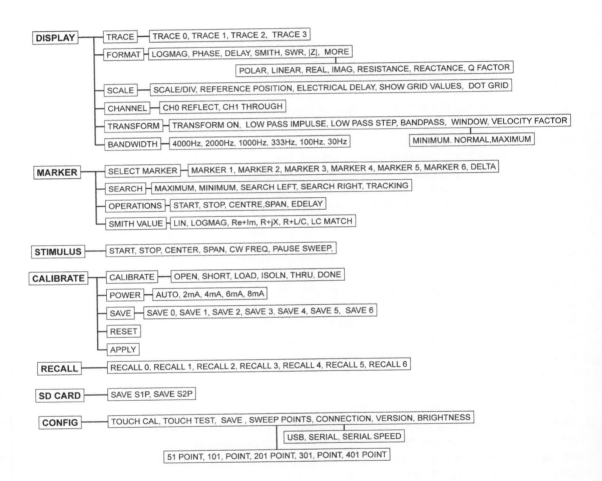

# Menu Maps

## NanoVNA V2 plus 4 - Menu Map

This menu map is based on firmware version: git-20201013-32077fd

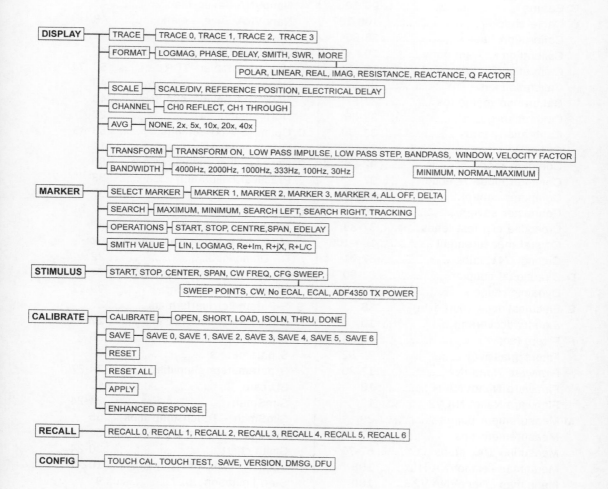

## Index – Alphabetical

**A** Active filter / amplifier ................. 11, 82
ADF4350. ..................................... 15
Analysing RF networks .............. 25
Antenna connections ................. 59
Antenna feeder loss ................... 65-67
Antenna measurement .............. 58-63
Attenuator measurement ........... 85
ATU settings - optimise ............. 63-65
**B** Baluns ......................................... 96-100
**C** Cable checker ............................ 106-107
Calibration basics ...................... 28-30
Calibration kit - 3rd party ........... 29
Calibration kit - supplied ........... 28, 32
Calibration kits - DIY ................. 36-37
Calibration memories ................ 32
Calibration guide ....................... 33-36
Calibration status ...................... 33, 39
Cartesian chart .......................... 21
Common mode chokes ............. 92-96
Common mode rejection .......... 97
Computer control ....................... 40-53
Connector savers ...................... 30
Crocodile clip test leads ............ 37-38
Crystal measurement ................ 104-106
Cutting 1/4λ stubs ...................... 67-71
**D** Directional couplers .................. 87-90
Dynamic range ........................... 10, 78
**E** Electrical delay (port extn) ........ 38
Extendedcoverage ..................... 10
**F** Filters (active) ............................. 82-85
Filters (passive) .......................... 80-82
Firmware NanoVNA ................... 17-20
Firmware NanoVNA-H .............. 18
Firmware NanoVNA V2 ............. 19
**M** Measurement points ................. 29
Measurement tips ..................... 57-58
Measuring 1/4λ stubs ............... 67-71
Menu map - NanoVNA-H........... 108
Menu map - NanoVNA V2 ......... 109
**N** NanoVNA basics ....................... 7-9
NanoVNA original design ......... 14
NanoVNA quirks ........................ 9-13
NanoVNA V2 .............................. 9, 15-16
NanoVNA V3 .............................. 16
NanoVNA versions .................... 13-17
NanoVNA Web app .................... 51

NanoVNA Web app ..................... 51
NanoVNA-F ................................. 15
NanoVNA-H ................................. 11, 14
NanoVNA-QT ............................... 20
NanoVNA-Saver .......................... 9, 40
NanoVNA-Saver - analysis ......... 49
NanoVNA-Saver - data analysis... 47
NanoVNA-Saver - Linux ............. 40
NanoVNA-Saver - Mac ................ 41
NanoVNA-Saver - markers ......... 48
NanoVNA-Saver - Raspberry Pi .. 41
NanoVNA-Saver - sweep panel .. 45
NanoVNA-Saver - TDR ............... 50, 71
NanoVNA-Saver - Windows ....... 40
NanoVNA-Saver - segments ...... 43
Normalising ................................. 27
**O** Optimise ATU settings ............... 63-65
Output level adjustment ............. 11
**P** Passive filter measurement ...... 80
Practical calibration ................... 33
Practical measurements ........... 57
**Q** Quarter wave stubs .................... 67-71
**R** Reflectance bridge ..................... 8
Measure antenna from shack ... 62
RF Demo kit ................................ 22
RF switch & relay measurement .. 74
RF tap measurement ................. 90-92
**S** S Parameter file formats ........... 27
S parameter normalising .......... 27
S parameter numbering ............ 26
S parameters .............................. 7
S parameters simplified ............ 24-27
SD card ........................................ 11
SimSmith ..................................... 24-24
SimSmith - Touchstone files .... 55
SimSmith and NanoVNA ........... 54-56
Smith chart ................................. 20-24
Splitter / combiner ..................... 102-103
Step limitation ............................ 9
**T** Test cables ................................. 31
TDR .............................................. 50-51, 71-74
Touchstone files ......................... 11
**U** Ununs .......................................... 100-102
**V** V2 crystal measurements ......... 13
VNA measurements ................... 25

# Index – by Subject

**Amplifiers**
    Active filter / Amplifier ............ 11, 82-85
**Antennas**
    ATU settings ..................................... 63-65
    Connections ........................................... 59
    Crocodile clip test leads ................... 38
    Cutting 1/4λ stubs ............................ 67-71
    Feeder loss ........................................ 65-67
    General measurement .................... 58-62
    Measurement from shack ............. 62-63
**Attenuators**
    20dB home-brew attenuator .......... 78
    Measurement ..................................... 85-87
**Calibration**
    3rd party calibration kits ................ 29-30
    Basics ................................................. 28-39
    Connector savers ............................ 30
    Homebrew calibration kits ............. 36-38
    Memories ........................................... 32-33
    Practical calibration example ......... 33-36
    Supplied calibration kit ................... 28, 32
**Filters**
    Active filter measurement ............. 11, 82-85
    Passive filter measurement ........... 80-82
**Firmware**
    NanoVNA V2 ..................................... 19-20
    NanoVNA-H ...................................... 17-19
**Menu Maps**
    NanoVNA V2 ..................................... 109
    NanoVNA-H ...................................... 108
**NanoVNA Background**
    NanoVNA V2 block diagram ........... 15
    NanoVNA block diagram ................ 13
    Quirks ................................................. 9
    Versions ............................................. 13-17
**Practical Examples**
    Active filters and amplifiers ............ 82
    Antennas ............................................ 58
    Attenuators ....................................... 85
    ATU settings ..................................... 63
    Baluns ................................................ 96
    Cable checker .................................. 106
    Common mode choke ..................... 92
    Crystals ............................................. 104

    Cutting 1/4λ stubs .......................... 67
    Directional couplers ........................ 87
    Feeder loss ....................................... 65
    Passive filters .................................. 80
    RF tap ................................................ 90
    Splitter / combiner .......................... 101
    Switches and relays ........................ 74
    TDR - distance to fault ................... 71
    Ununs ................................................ 100
**Reflectance bridge** ............................ 8
**Smith chart**
    Introduction ...................................... 20-24
    SimSmith ........................................... 24
    SimSmith - Touchstone files ......... 55-56
    SimSmith with NanoVNA ............... 54-56
**Software**
    Computer control ............................ 40-54
    NanoVNA-Saver ............................... 9, 40-51
    NanoVNA web app .......................... 51
    NanoVNA-Q ...................................... 20
    NanoVNA-Saver - Analysis modes  49
    NanoVNA-Saver - Data analysis .... 47
    NanoVNA-Saver - Display config ... 45
    NanoVNA-Saver - Sweep panel ..... 45
    NanoVNA-Saver - Segments .......... 42
    NanoVNA-Saver Installation - iOS.. 41
    NanoVNA-Saver Installation - Linux..40
    NanoVNA-Saver Installation - Pi ..... 41
    NanoVNA-Saver Installation - Win....40
**Test leads**
    Antenna connections ...................... 59
    Cable selection ................................ 31
    Connector savers ............................ 30
    Crocodile clip test leads ................. 38
    Port extensions ................................ 38
**Time Domain Reflectometry**
    Introduction ...................................... 71-74
    W2AEW formula and table .............. 72
**Touchstone files**
    NanoVNA -Saver file format ........... 47-48
    NanoVNA with SimSmith ............... 54-55
    NanoVNA-Saver - exporting ........... 51
    S-parameter file format ................. 27

## Summary and Resources

I hope you have found this book to be a useful reference for taking measurements with the surprisingly versatile NanoVNA. If you would like to discover even more ways to use your NanoVNA, please check the video and tutorial links shown below.

### User Groups

*NanoVNA Explained:*  https://groups.io/g/nanoVNA-Explained
*Original and -H NanoVNA:*  https://groups.io/g/nanovna-users/
*NanoVNA V2:*  https://groups.io/g/NanoVNAV2
*SimSmith:*  https://groups.io/g/SimSmith

### Software

*NanoVNA-Saver:*
   https://github.com/NanoVNA-Saver/nanovna-saver/releases
*NanoVNA-App:*
   https://nanovna.com/?page_id=141
*NanoVNA-QT:*
   https://github.com/nanovna-v2/NanoVNA-QT
*SimSmith*:
   https://www.w0qe.com/SimSmith.html

### Home Websites

*NanoVNA:*  https://nanovna.com/
*NanoVNA V2:*  https://nanorfe.com/nanovna-v2.html

### Video Tutorials

*W2AEW - Lots of NanoVNA tutorials:*
   https://www.youtube.com/user/w2aew/videos
*W0QE - Excellent SimSmith tutorial series:*
   https://www.w0qe.com/SimSmith.html

### Reference Material

*Rohde & Schwarz:*
   https://www.signalintegrityjournal.com/ext/resources/White-papers-App-notes/Vector-Network-Analyzer-Fundamentals-Primer.pdf
*Keysight:*
   https://www.keysight.com/gb/en/assets/7018-06841/application-notes/5965-7707.pdf

### Calibration kits

*SDR-Kits:*  https://www.sdr-kits.net/